Longevity and the Good Life

Longevity and the Good Life

Anthony Farrant

palgrave
macmillan

First published 2011 by
PALGRAVE MACMILLAN

Palgrave Macmillan in the UK is an imprint of Macmillan Publishers Limited, registered in England, company number 785998, of Houndmills, Basingstoke, Hampshire RG21 6XS.

Palgrave Macmillan in the US is a division of St Martin's Press LLC, 175 Fifth Avenue, New York, NY 10010.

Palgrave Macmillan is the global academic imprint of the above companies and has companies and representatives throughout the world.

Palgrave® and Macmillan® are registered trademarks in the United States, the United Kingdom, Europe and other countries.

ISBN 978–0–230–57695–7   hardback

This book is printed on paper suitable for recycling and made from fully managed and sustained forest sources. Logging, pulping and manufacturing processes are expected to conform to the environmental regulations of the country of origin.

A catalogue record for this book is available from the British Library.

Library of Congress Cataloging in Publication Data

Farrant, Anthony.
    Longevity and the good life / Anthony Farrant.
        p. cm.
    Includes bibliographical references and index.
    ISBN 978–0–230–57695–7 (alk. paper)
    1. Longevity—Philosophy.   2. Longevity—Moral and ethical aspects.
I. Title.
    RA776.75.F35 2010
    305.26—dc22

                                                                    2010033944

10   9   8   7   6   5   4   3   2   1
20   19   18   17   16   15   14   13   12   11

Printed and bound in Great Britain by
CPI Antony Rowe, Chippenham and Eastbourne

#6572270859

Contents

Acknowledgements   vii

Introduction   viii

1   Longevity, Technology and Humanistic Values   1
The human life span   1
Enhancing life spans   12
Technology and humanistic values   15

2   The Misfortune of Death   26
Longevity and the misfortune of death   28
Epicurean and Lucretian objections   31
The phenomenological problem of death   32
The ontological problem   34
The asymmetry of non-existence   40
The tedium of immortality   44

3   Justifying the Means   50
Animal experimentation   51
Embryo experimentation   60

4   Longevity and the Problem of Overpopulation   73
Preliminary aspects of population theory   73
Carter's critique of Parfit   79
The problem of increasing longevity   84

5   Ending Lives   87
The fair innings argument   88
Limiting life spans   95
Limiting longevity, suicide and euthanasia   99

6   Partiality and Equality   113
The principle of equality   113
Establishing equality   117
Partialism   120
Human nature and equal respect   125
Equality and longevity   134

7 Longevity and the Good Life   **139**

Notes   145

Bibliography   156

Index   165

Acknowledgements

I have benefited greatly from the advice, guidance and patience of Suzanne Uniacke. I also owe a debt of gratitude to Brenda Almond. I have profited from early discussions on this topic with Piers Benn, Stephen Burwood, Paul Gilbert, Stella Gonzalez-Arnal, Ashley Harrold and Constantine Sandis. I am grateful to Stephen Holland and Tom Stoneham in the Department of Philosophy at the University of York for their support of my research. Finally, I am particularly indebted to my parents for their generous support and to Claudia Capancioni for her encouragement and understanding.

I have drawn on my earlier paper, 'The Fair Innings Argument and Increasing Life Spans', (2009), *Journal of Medical Ethics*, 35, 53–6, for part of my argument in Chapter 5. I am grateful to an anonymous referee for the *Journal* for his or her insightful comments.

Introduction

Human beings, as with most other animals, appear to have a biologically limited life span. There is no reasonable evidence of any human being having lived much longer than 120 years, and relatively few people have achieved such a long span of life. The search for a way to increase longevity, along with a concomitant prolongation of the quality of health, has a long history and cultural presence. The *Epic of Gilgamesh* may be the earliest tale of a search for a means to prolong life while the idea of a 'fountain of youth' is an old, culturally diverse and widespread myth.[1] Gerontological knowledge about why human beings age and the implications of this for restricting how long we can live improved considerably over the course of the last century, and continues to develop. Along with the potential of biotechnologies, this knowledge suggests it could now be possible to enable people to live longer, healthier lives.

Medical interventions that enhance human functions and form raise considerable anxieties, and endeavours to increase life spans are no different. The unease felt about biotechnologies concerning, for example, the effects of their use on human nature, can be placed within wider concerns about modern technology and their impact in shaping and determining our values. Longevity may be valuable, but the very real prospect that we could increase life spans raises questions about the role of technology in distorting our values. The search for longer life spans may have an extensive history, but it is not an endeavour that has gone unopposed. It is from within the context of this wider debate that I consider some of the ethical implications of increasing life spans for the pursuit of the good life.

Throughout, I take a humanist approach, and broadly restrict my focus to the way in which life extension affects our moral relationships with others. I include among these others nonhuman animals for reasons that will become more obvious as the argument progresses. Nevertheless, my inclusion of nonhuman animals points to a pervasive theme in my argument about the search for longer, healthier lives as a humanist endeavour.

Central to my assessment is the idea that the good life is fragile, that it is subject to the contingencies of life, which in turn shape and contribute to our values. The extension of life spans beyond their apparent biological limit represents an attempt to control one of the fundamental vulnerabilities of life, namely, death and the uncertainty of when it will occur. To be successful at increasing longevity will demonstrate a triumph of reason over the non-rational aspects of life, the contingencies that defy control and destabilise the pursuit of the good life. In so doing, it will give the impression of further distinguishing human beings from the natural world, including nonhuman animals. To conceive of increases in longevity in this way is not an understanding that I will endorse, but it will be an underlying theme throughout my argument.

The idea that prolonging life spans is to control death points to another constraint on my argument. My concern is with increases in life spans and, by its very nature, this does not involve the attempt to make human beings immortal. This constraint exists for two reasons. First, even to remove completely the biological causes of the limitations on human longevity would not make people immune to death. Second, I take the positive view that advances in medicine will be such that people will live longer, healthier lives. Nonetheless, any increases in life spans will be gradual and there is no prospect in the immediate future, if at all, of a significant extension to longevity such that people might live for hundreds of years.[2]

There are three final caveats concerning my argument. In taking a humanist approach, I restrict my analysis to the implications of increasing longevity on Western values; it may be the case that in different cultures some of the issues I discuss will not arise. Furthermore, I make the assumption that any increases in life spans will apply to every individual in society. In practice, this may not be the case, but a crucial aspect of my argument concerns the consequences of extending life spans on the idea that human beings are fundamentally equal. In this regard, I do not address the practical issues as such of prolonging life; my aim is to isolate the key values that increasing life spans will affect. Finally, as may be obvious from the title of this book, I approach the issues from a non-consequentialist perspective on ethics.

1
Longevity, Technology and Humanistic Values

Advances in biotechnologies and developments in the knowledge of the biology of ageing mean it may soon be possible to increase human life spans. It is already the case that in the West, average life expectancy increases by approximately five hours a day (Kirkwood, 2008, p. 644). What the potential of biomedicine promises, however, is the possibility of increasing longevity beyond the apparent biological limit to how long we can live. The use of technology to enhance people in this way raises a number of ethical concerns. The purpose of this first chapter is twofold. The first is to outline three broad areas for assessing what these concerns are, which will structure the argument of the book. The second, and in relation to the first area of assessment, is to consider the relationship between technology and our values.

The human life span

Before any assessment of the ethical implications of increasing life spans can begin, it is essential to be clear about how long we can and do live. The fact that human beings do not live for hundreds of years implies there is a biological limit to human longevity, as with many other animals. What follows is an outline of the principal theories explaining why such a limit exists and how it functions.

The length of life

The traditional, biblical account of the human life span maintains that we can expect to life for 70 years.[1] Until recently, average expected life spans in the West did not differ much from this tradition, with average expected life spans at birth now being approximately 80 years. Nevertheless, there are many examples of people living longer than this. Sophocles is reputed to have died aged 90, and the number of people reaching their 100th year in the UK numbers in the thousands (Kirkwood, 2000, p. 6). The oldest known person was Jeanne Calment, who died aged 122 years and 164 days. While there have been many people who have claimed to be older than Calment, there is insufficient evidence to support their claims.

Although the present average expected life span tends to reflect the traditional account of the human life span, this is a relatively new phenomenon. Average life spans in ancient Greece were approximately 30 years (Austad, 1997, p. 32), and even by the end of the nineteenth century, average life spans were approximately 46 years. Yet, by the end of the twentieth century, average life expectancy had almost doubled.[2] This doubling of average life expectancy was in part the result of improvements in sanitation, diet, social conditions, and the development of antibiotics and vaccinations against infectious diseases. Of greater significance is the fact that the measure of life expectancy is statistical. The average expected life span at birth is a statistical estimate of how long the newly born of a particular year might expect to live based on the average age of mortality for the same year. The doubling of life expectancy over the course of the twentieth century was greatly influenced by significant reductions in infant mortality rates. For example, in England and Wales in the 1880s, only 74 per cent of infants survived until their fifth birthday. By the 1990s, this rate had increased to over 99 per cent (Kirkwood, 2000, pp. 5–6).

The average expected life span at birth provides a statistical estimate for how long people of a particular birth cohort can reasonably expect to live, given their social, cultural, economic and environmental conditions. It does not indicate for how long individuals could live: it does not identify a biological limit to human longevity. The example of Jeanne Calment is clear evidence that, as a species, human beings are capable of living longer than the present life expectancy. Indeed, to date, the apparent maximum life span for human beings is approximately 120 years, but this is determined by the documented oldest-lived person and not by any biological limit (Olshansky et al., 2001, p. 1491).

The question then arises as to whether there is a biological limit to how long human beings can live. Bruce Carnes et al. (2003, p. 32) refer to a 'biological warranty period', which identifies the species typical biological limits to longevity. The concept of a biological warranty period identifies the typical intrinsic capability of the organism of a particular species to sustain life. It might be compared to the warranty period of a machine. All machines of a certain type will be constructed to last for a certain number of years. Some machines will not last for as long as the warranty period for a number of reasons to do with internal faults and external factors; nevertheless, some machines will also survive for longer than the warranty period. The biology of the human organism is intrinsically capable of sustaining life for only a certain period of time. As with machines, many human beings will fail to live for the length of the warranty period for a variety of reasons, and some will live longer than it.[3]

Carnes et al. (2003, p. 41) estimate the biological warranty period of human beings to be an average expected life span of 85–95 years. To achieve an average expected life span at birth of 85 years, S. Jay Olshansky et al. (1990, 2001) calculate that mortality rates would have to be reduced by 50 per cent for every age group. Further increases in average expected life spans at birth might be possible, but each yearly increase would require progressively larger reductions in mortality for every age group. Given the present level of developments in biomedicine and extrinsic causes of death, such as accidents, homicide and suicide, it is unlikely that average life expectancy will increase to beyond the 85–95 year biological warranty period that is claimed is typical for human beings (Olshansky et al., 2001, p. 1491; Carnes et al., 2003, p. 41). To achieve increases in longevity beyond this limit, Olshansky et al. (2001) argue it will be necessary to add years to the lives of people who reach their 70s and 80s by combating the intrinsic causes of the warranty period.

Ageing and the limits of longevity

It will only be possible to achieve the addition of extra years to the lives of those who reach their 70s and 80s by understanding why and how the biological warranty period works. The intrinsic biological mechanism that limits longevity is the ageing process, where ageing can be defined as the increasing susceptibility of individuals as they grow older to intrinsic or extrinsic factors that may cause their death.[4] It has been calculated that the probability of mortality for modern human beings in the West doubles every eight years. Similar patterns in the doubling of the probability of mortality are detectable in other animal species, although the rate of doubling differs between species.[5] There have been a number of theories about why animals age, from which at least three competing ideas are discernible.

The inevitability of ageing

One theory maintains that the ageing process is the inevitable result of the gradual deterioration that our bodies experience over time. What restricts the biological warranty period of a machine is its ability to cope with the gradual wear and tear that it experiences as a result of its daily functions and environment. The brakes on a motor car, for instance, will gradually wear down with continued use until they no longer function or are repaired. The ageing process, according to this theory, is much the same for animals.

For example, oxygen is essential for human beings to live, but its consumption is eventually harmful to us. Oxygen reacts with other chemicals, which makes it useful because it is used to break down fats and carbohydrates and in so doing produce energy. This feature of oxygen is also harmful because it will react with other types of molecule to produce free radicals, which are unstable molecules that can react with other stable molecules to produce more unstable molecules. The result causes damage to DNA, proteins, and ultimately cells (Austad, 1997, p. 84). Over time the gradual damage caused by oxygen free radicals accumulates, which along with other types of damage, such as mistakes in the replication of DNA during cell division, lead to the increasing susceptibility to intrinsic and extrinsic factors that may cause death.

In criticising the idea that ageing is the inevitable consequence of wear and tear, Tom Kirkwood (2000, pp. 53–4) observes that it relies upon the second law of thermodynamics, which maintains that in a closed system entropy increases. In accordance with this law, the wear and tear of our bodies develops disproportionately as we grow older, thereby explaining our greater probability of mortality as we age. Unlike machines, however, human beings are not closed systems and our bodies take in and expel elements from the environment to obtain energy, which can be used to combat entropy. Indeed, our bodies possess mechanisms that utilise this energy for repairing the damage that we experience. Furthermore, all animals, just as with all machines, will experience gradual wear and tear as a result of existing and functioning, but not all animals experience ageing. While the fact that we experience wear and tear might explain how we age, it cannot by itself explain why ageing occurs (Kirkwood and Austad, 2000, p. 233).

Ageing is programmed

Another theory that purports to explain why ageing occurs maintains that it is the result of a necessary programme. According to this theory, specific genes have evolved to limit the typical life span of any member of a particular species in order to prevent overpopulation (Kirkwood, 1997, p. 1766; 2000, pp. 59–60). If ageing is the result of a genetically controlled programme, this might explain why different animal species age at different rates (Kenyon, 1996). There are a number of biological mechanisms that are alleged to be evidence of such a programme. The Hayflick Limit is one example (Kirkwood, 2000, p. 91). Leonard Hayflick discovered in the 1960s that soma cells are capable of only a certain number of cell divisions. Once a cell reaches its Hayflick Limit – the maximum number of times it can divide – it becomes senescent.

The idea that ageing is programmed in order to limit longevity and thereby prevent overpopulation is untenable. In contrast to human beings, animals do not generally live in protected environments and do not, on the whole, live long enough to experience ageing (Medawar, 1952, p. 13; Kirkwood, 1997, p. 1766; 2000, pp. 60–2). It is only because of exceptions, and where individuals are reared in controlled environments, that the evidence of ageing among different animal species becomes apparent. As a consequence, ageing does not appear to be necessary for controlling population levels. What is more, evolutionary theory does not support the idea of a programmed limit to longevity because evolutionary adaptations favour individuals and not their species. In a species where life spans are programmed, a mutation that allows individuals to live longer than is typical of the species is advantageous. A longer than normal life span enables individuals to have access to more resources and also to have a larger number of offspring. As a result, over time the mutation would spread until longer life spans became typical for the species. The genes that limit longevity would evolve out of existence. While a shorter life span might benefit the species and prevent overpopulation, so long as longer life spans benefit individuals, a gene for shorter life spans will not re-evolve (Austad, 1997, pp. 55–62; Kirkwood, 2000, pp. 60–2).

The Disposable Soma Theory

Despite their flaws, theories about the inevitability of ageing and a programmed limit to longevity point to a more successful account of how and why we age. A theory of ageing must explain why animals that experience ageing are unable to repair the wear and tear they suffer, and this explanation must satisfy the demands of evolutionary theory. Three evolutionary based theories are discernible, although Kirkwood and Steven Austad (2000, p. 233) view them as complementary rather than conflicting ideas. Of these, Kirkwood's own Disposable Soma Theory is the more compelling.

The role of natural selection

Two early theories focus on the role of natural selection. The first, now called the Mutation Accumulation Theory, derives from Peter

Medawar's observation (1952) that the influence of natural selection on individuals weakens as they grow older.[6] Harmful mutations that express themselves early in life may prevent organisms from reproducing, which means they will not be expressed in future generations. Late-acting, harmful mutations may be expressed in future generations, but their harmful effects are less important, both to individuals and in terms of evolutionary fitness. As late-affecting mutations, they will not interfere with early reproduction, where reproductive success indicates evolutionary fitness, and extrinsic sources of mortality mean individuals are unlikely to survive long enough to experience their harmful effects. The effects of these mutations, should any individual live sufficiently long to experience them, will be expressed as ageing.

A second theory, developed by George Williams (1957), concerns the effects of pleiotropic genes. The Antagonistic Pleiotropic Theory postulates the idea that there may be genes that have good effects early in life but harmful effects later in life. Williams provides the theoretical example of a gene that deposits calcium in bones during early life but with the consequence that this leads to a hardening of the arteries in later life (p. 402; Kirkwood, 2000, p. 76, also uses the same example). Pleiotropic genes are beneficial when extrinsic forces limit life spans such that individuals will not normally live long enough to experience their harmful effects. Once again, individuals who live longer than the normal species typical life span will experience the harmful effects of pleiotropic genes as ageing.

The Disposable Soma Theory

The two theories of Mutation Accumulation and Pleiotropy provide important insights into why ageing occurs, but there is lack of evidence to support such processes or for pleiotropic genes (Kirkwood, 2005, p. 438). In contrast to the earlier theories, Kirkwood's Disposable Soma Theory does not identify specific genetic functions but points to the physiological process of ageing in terms of trade-offs (Kirkwood, 2005, p. 439). Kirkwood draws on the idea that our bodies experience wear and tear and provides an evolutionary explanation to show why this is allowed to occur.[7]

Kirkwood's argument develops upon the ideas of August Weismann, who argued that there is a distinction between soma cells and the germ-line and that the germ-line must be immortal.[8] Genes can continue in existence either through the indefinite survival of individuals or through reproduction. The chances of an individual surviving indefinitely are negligible because they are subject to extrinsic factors, such as diseases and accidents, which will eventually cause death. As a result, if genes are to survive it makes sense for individuals to reproduce.

A problem arises, however, regarding how best to utilise the available resources. Much of the energy we obtain is used for basic biological functions, such as metabolism and the maintenance of soma cells. For the most part, resources are scarce meaning individuals struggle to obtain sufficient energy for their basic needs, but even where resources are not scarce there remains a limit to how quickly they can be utilised (Kirkwood and Rose, 1991, p. 16). Resources are also required for reproduction, and while this provides the best means of survival for our genes, individuals must live long enough to reproduce. This creates a predicament about how best to allocate resources, because if they are used for maintenance they cannot also be used for reproduction. The evolutionary solution to this problem is a trade-off between soma cell maintenance and reproduction.

The trade-off involves favouring reproduction at the expense of soma cell maintenance. In order to ensure that enough energy is made available for reproduction, soma cells receive insufficient energy to prevent the accumulation of damage to them. The energy they receive is sufficient for them to maintain an individual for the duration of the life span that is typical for their species, as is determined by extrinsic causes of mortality. This is necessary if an individual is to maximise their opportunities for reproduction, but to repair and maintain soma cells so that an individual can live longer than the species typical life span is an inefficient use of energy. As a result, an individual experiences the effects of the gradual accrual of damage as ageing when they live longer than is typical for their species.

The need to invest in bodily maintenance early in life and the existence of a trade-off of in energy between maintenance and reproduction is indicative of the idea of pleiotropy. For instance, damaged cells can become dangerous to organs. A theoretical way of resolving this is for there to be a mechanism shutting down damaged cells. This would work well early in life, but as an organism grows older it is likely to experience an increasing number of damaged cells, which if shut down would, because of the large numbers involved, compromise the functioning of organs and be recognisable as ageing (Kirkwood, 2005, p. 442). Whether such mechanisms exist is unclear, but the idea of pleiotropy captures the essence of the Disposable Soma Theory.

Two final points should be noted about the Disposable Soma Theory. The first is that it points to the now widely accepted view that ageing is caused by the accumulation of molecular and cellular damage

(Kirkwood, 2008, p. 645). Nevertheless, the mechanisms of the ageing process are complex and still to be understood. If human life spans are to be increased, more research on and knowledge about what causes this damage and how it affects cells and tissues is needed. Study of the ageing process also seems relevant and necessary for the development of treatments for age-related diseases such as cancer, heart disease and dementia because they, like ageing, are often the result of damage to cells (p. 646).[9] The second point to note about the Disposable Soma Theory is that it can explain why life spans vary between animal species and for individuals of a particular species. The effects of ageing become noticeable when an organism lives longer than the typical life span for its species, given the constraints of extrinsic factors. The constraints will be specific to a particular species and determine how much energy should be provided for soma cell maintenance (Kirkwood and Austad, 2000, p. 233). What is more, the accumulation of molecular and cellular damage will vary between individuals, in part because of social and environmental factors, thereby explaining why people, for example, have different life spans.

Kirkwood argues that human beings experience ageing because human evolutionary biology has not kept pace with the evolution of culture and society (Kirkwood, 2000, p. 80). The fact that we experience ageing is because our bodies evolved in response to greater extrinsic causes of mortality than we now encounter. For example, the doubling of the average expected life span during the twentieth century was because of modest advances in medicine and improvements in social conditions, which, by controlling our environment, reduced the extrinsic causes of death. As Carnes et al. (2003) observe, it might be possible to increase average life spans further by reducing mortality caused by extrinsic factors, but any increase beyond the biological warranty period of approximately 85–95 years must address the causes of ageing, namely the gradual accumulation of damage at the molecular and cellular level. Advances in gerontology and biotechnology mean the evolutionary need to trade cell maintenance for improved reproductive prospects may in the future no longer be necessary. Human intervention could ensure both are possible without the requirement of a compromise.

Ageing as a disease

In arguing that the biological evolution of human beings has not kept pace with our cultural and social evolution, Kirkwood (2000) concludes that ageing is neither necessary nor inevitable. To know why people age is a necessary first step to knowing how people age and what is needed to prevent or slow down the ageing process. Indeed, the increasing knowledge of why and how people age has led some scientists, such as Leonard Guarente and Cynthia Kenyon (2000), to argue that ageing is a disease, although others, such as Hayflick (2000), deny this.[10] These conflicting claims depend upon what constitutes a disease. It is beyond the aims of this discussion to consider this concept in detail, but the claim that ageing is a disease along with the evaluative nature of this concept will help to illustrate my argument as it progresses.

The concept of disease is used as the antonym of health and identifies any unhealthy condition (which, Norman Daniels (2008, p. 36) observes, is in contrast to the ordinary usage of the term 'disease'). Christopher Boorse (1975) argues that the traditional account of health is that it is an evaluative concept because to be healthy is valuable. Yet Boorse denies this because he claims we can distinguish between health as a 'descriptively definable condition' and valuing this condition (p. 54). He defines a healthy condition as the normal species typical functioning of an organism. In so doing, Boorse (p. 57) draws upon C. Daly King's account (1945) of normal functioning as referring to the natural design of an organism's functions.[11] Thus plants, for instance, have naturally developed to photosynthesise, and any condition that prevents this process is a diseased state. Health, therefore, is normal species typical functioning, and it is this condition that we value.

A particular difficulty for any definition of health based on normal functioning is identifying what constitutes a normal function. This is particularly important for the idea that ageing is a disease (and for the enhancement debate, which I consider below). The Disposable Soma Theory maintains that ageing, the gradual accumulation of damage at the molecular and cellular level, is the result of an evolutionary trade-off. From an evolutionary perspective, the normal functioning of a cell is to degrade over time. The Disposable Soma Theory provides what I will refer to as a teleological account of the normal functioning of soma cells. In contrast, the idea that ageing is a disease denies that ageing is normal. It identifies the normal functioning of soma cells as their routine processes, which the ageing process prevents them from undertaking. The claim that ageing is a disease rests on what can be called a practical account of the normal functioning of cells.

To identify the normal functioning of soma cells as its natural design will not necessarily indicate which account is correct. The natural design may be teleological or it may be practical. Both types of function identify processes that are inherent to the cells, which might provide a criterion for a design being natural. An opponent of the idea that ageing is a disease might point to this criterion and observe that the practical functioning of soma cells is sustainable only with human intervention, in contrast to the teleological function. Such a response implies that for a function to accord with its natural design it cannot be influenced by human activities. To argue for this requires that a clear boundary can be drawn around human activities and where their influence on the world ceases, and it is not obvious that this is possible. What is more, the response implies that human beings are not part of nature; if we are, then there will be no distinction between nature and human activities.[12] If the opponent accepts that human activities can influence our natural design to some extent, they must explain why interventions that sustain the practical function of cells are unacceptable. There may be a justifiable explanation, but it will be dependent upon a normative judgement.

It may be the case that some functions and accounts of natural design are purely descriptive. Where there are rival accounts, however, as there are with the normal functioning of soma cells, the decision about what function is normal and which design is natural will be normative. As such, it is not always possible, and is impossible in the case of health, to draw a distinction between a description of a function and valuing this function because the value of the function will determine whether it is normal. What motivates the view that ageing is a disease is the undesirability of ageing, specifically, the decline in the quality of health that this involves.[13] And it is the value we place on a good quality of health which determines that the practical account of the functioning of soma cells is normal. This choice is also influenced by the fact (and the belief) that human beings can and do intervene to maintain or restore the quality of health.

The length of life and the quality of health

There are two ways of conceiving of increases in life spans. They may be looked upon as the incidental consequence of medical endeavours to maintain and restore good health; or be regarded as the successful result of intentional endeavours to increase longevity. There is, I propose, little substantive difference between these approaches, and explaining why serves a number of illustrative purposes.

For the most part, increases in life spans will be an incidental consequence of maintaining or restoring the quality of health. The ageing process gives rise to greater frailty and vulnerability as we grow older; and associated with this process and older age groups are a number of diseases and disorders, such as Alzheimer's disease, Parkinson's disease, stroke, cancer and cardio-vascular disease. To be able to treat or cure ageing and the diseases and disorders associated with it will lead to a better quality of health as we grow older, but it will also have the consequence of prolonging life. Olshansky et al. (2001) point to the need to reduce mortality rates for each age group and to add extra years to the lives of those who reach their 70s and 80s if life spans are to increase. But this is precisely what will be achieved if advances in medicine are such that survival rates for cancer increase, for example, and the vulnerability of ageing is reduced. The description of increases in longevity as an incidental consequence of maintaining and restoring the quality of health is disingenuous. The treatment of cancer serves not only to restore good health but also to prolong life. The same is true of many other medicines, which aim at restoring health and thereby preventing death, although not all medicines have this aim.

A second way of viewing increases in life spans follows a tradition with a long history, namely the search for a mythical 'elixir' of youth that can prolong life. In this case, any increase in life spans will be the result of endeavours that specifically intend such an outcome. If the Disposable Soma Theory is correct, it is unlikely that a single medicine could achieve longer life spans. Rather, any intentional efforts to prolong life will have to address the ageing process directly by preventing and repairing the many causes of molecular and cellular damage, as well as treat or cure the diseases and disorders associated with ageing. So doing would have the beneficial consequence of also maintaining good health (excluding extrinsic sources of poor health) while prolonging life. Most people would not want the fate of Tithonus, who was granted immortality but not eternal youth and withered as he aged without dying.

What these two conceptions of increasing life spans indicate is that there is a close relationship between good health and longevity. The relationship is, I propose, more appropriately between longevity and the quality of life. Health is instrumentally valuable because it is a means for us to possess those things that make life good. It is important to note that even with poor health life may be of sufficient quality for a person to want to continue living; and even with good health, life may be of insufficient quality to want to continue living. As the relationship between the qualities of health and life suggests, longevity is also valuable. It is not valuable for its own sake, but is structurally part of the quality of life: when life is good we want it to continue. I elaborate upon this claim in the following chapter. My aim here is simply to emphasise the value of longevity and how it relates to good health.

Two points must be emphasised. First, both contrasting approaches to longevity would appear to support the practical account of what constitutes normal functioning for soma cells. Increases in life expectancy beyond the biological warranty period, which are apparently incidental, will mean biological evolution has not kept pace with cultural evolution, implying that the teleological account would no longer be normal. Attempts at intentionally increasing life spans tacitly support the idea that ageing is abnormal and therefore a disease. I will consider further the role of these accounts of the normal functioning of soma cells in relation to the length of life in Chapter 5. Second, whether greater longevity is 'incidental' or intentional, the method for obtaining it, namely preventing or treating the ageing process and its associated diseases and disorders, means any increases are likely to be gradual. There is much more to learn about the mechanisms of the ageing process before any biomedicines can be developed to respond to ageing.

Enhancing life spans

The Disposable Soma Theory provides an explanation for the intrinsic limit to human longevity. How long we can live is restricted by the ageing process, and any medical interventions that can slow down or prevent it and its associated diseases and disorders will lead to an increase in life spans. As such, extending average life spans beyond the biological warranty period appears to constitute a form of medical enhancement.[14] In contrast, the idea that ageing is a disease implies that increases in life spans are the result of normal medical practices that maintain and restore the quality of health.

There is much debate about what constitutes a medical enhancement and why they are, if indeed they are, objectionable.[15] In what follows, I will define what I mean by a medical enhancement and outline three examples of enhancements that raise different concerns about them. The examples point to three different aspects of medical interventions that do or appear to enhance people that should be considered when assessing whether or not the interventions are morally permissible. Even if increasing life spans does not constitute a form of medical enhancement, the three areas of assessment will help to identify some of the issues that making people live longer will raise.

Medical enhancements are contrasted with medical treatments. Where treatments restore or maintain the human form and functions, enhancements improve them (Juengst, 1998, p. 31). As was noted earlier, an attempt to provide a non-evaluative account of health is to define it as normal species typical functioning. To enhance someone, therefore, is to augment their functions beyond the normal range that is typical for human beings.[16] A fundamental problem for defining health, and medical treatments, in this way is the widespread acceptance of preventative medicine. For example, vaccinations are used to prevent individuals from developing certain diseases, such as polio, but they work by enhancing our immune systems (Juengst, 1998, pp. 32–4). This suggests that either a new account of health is required or that the concept of enhancement is in part evaluative.

An alternative account identifies enhancements as medical interventions that do not meet 'genuine medical needs' (DeGrazia, 2005, p. 206). Daniels (2008, pp. 36–42) provides an account of health in terms of normal species typical functioning, but is prepared to allow a weak form of normativity in distinguishing between normal and abnormal functions. This enables his explanation of health and disease to accommodate some social accounts of disease, but only when they do not threaten the normal functioning account of health. Healthcare needs (genuine medical needs) are those medical and social interventions that promote, restore and maintain health over the course of one's life. By defining healthcare needs diachronically and by allowing, albeit in a weak way, a normative account of normal functioning, it can accommodate medical practices, such as preventative medicine, that might be construed as enhancements.

The difficulty for an account of genuine medical needs that is in part evaluative, no matter how 'weak' this normativity may be, is that it fails to clarify what enhancements are acceptable and what the moral concerns are that medical enhancements raise. Given the desirability of improving some biological forms and functions, what is needed is a further explanation identifying the reasons why particular interventions are objectionable. The following examples emphasise three principal concerns about medical enhancements.

Margaret Little (1998) uses the example of cosmetic surgery to highlight the way in which enhancement technologies might be used to promote unjust norms. Little observes that society will demonstrate preferences for certain types of physical appearance (pp. 164–7). In general, society is tolerant of those whose features do not meet one or more of its preferences. Nonetheless, some individuals who fail to meet a particular preference will seek cosmetic surgery to alter themselves, where their reasons for so doing are personal. This is not always the case, and some groups will be prejudiced against people because of their appearance, which might force those who experience this prejudice to seek cosmetic surgery. Little uses the example of a boy whose ears stick out (p. 164). Society may prefer flat ears and generally tolerates people whose ears stick out, but the boy is ridiculed by his fellow pupils. What is wrong with this situation is not the preference of society for flat ears or society's reaction to those whose ears do stick out but the prejudice of certain members of society. This contrasts with a situation where the preferences of society for a certain appearance are unjust and where there is a more general social pressure to conform to these preferences. For example, Little observes that women can experience social pressure to conform to certain visual ideals that do not reflect their natural or typical appearance (p. 167). Cosmetic surgery can reinforce these unjust prejudices by making it possible for people to conform to an unjust ideal of beauty.

A second example points to the way the means for achieving a certain objective can contribute to the objective's value. For example, Ritalin is used to improve the behaviour and attentive skills of children who have medically recognised attentive and behavioural disorders. Yet it might also be used to enhance the performance of children at school who do not have such disorders, by augmenting their attentive skills. While a good education is valued for its own sake, it is also valued because of the effort and sacrifices people must make to achieve it. The use of performance-enhancing psychopharmaceuticals, such as Ritalin, to obtain a good education undermines the value of educational success (Cole-Turner, 1998, p. 151).

A third example raises concerns about the consequences of medical enhancements. Carl Elliott (1998) argues that psychopharmacology can be used so as to prevent people from living authentic lives. Prozac is prescribed by physicians to stabilise or alter a patient's personality by making them less depressed. There are many situations that justify the prescription of Prozac for this purpose, but Elliott warns, a temporary state of angst or melancholia can also be beneficial. Existential angst can allow people to decide to live more authentic lives because it pro- poses that they should question or direct the way they live their lives. It can be added to Elliott's claims that temporary depression or angst might also serve as a cathartic response to the death of a loved one or the end of a relationship, which can help a person come to terms with these events. Elliott argues that using psychopharmacological drugs like Prozac risks damaging the values that are derived from fluctuations of specific dispositions. The vulnerability of our values that these fluctua- tions demonstrate contributes to the pursuit of the good life.[17]

These examples point to three areas that an assessment of the moral implications of medical enhancements might consider. The first addresses the values that a medical intervention promotes and how realising those values might distort them. This will be the focus of the remainder of this chapter and the one following. The second area considers how the means for achieving a particular end, such as good exam results, affects the value of successfully achieving this end, which I examine in Chapter 3. The final area assesses the consequences of the intervention and whether it is detrimental to the recipient of the intervention, to other people or both. This will primarily be the focus of the remaining chapters of the book.

Technology and humanistic values

Any increase in life spans will be dependent upon developments in gerontology that can provide more information about the mechanisms of the ageing process. Equally as important will be advances in biotechnologies that can prevent, repair and treat this process and the diseases and disorders associated with it. For the remainder of this chapter, I will focus on the relationship between technology and our values and anxieties that modern technology is shaping, distorting and determining them. The earlier account of cosmetic surgery provides an example of the distorting effect of technology that promotes one or more values to the detriment of others. My concern will be with the nature of modern technology and whether it is determining what values to promote by suggesting new objectives to pursue, where so doing could be disadvantageous to our ability to live a good life.

Earlier, I briefly claimed that longevity is valuable. If concerns about technology are correct, it could be that the pursuit of longer life spans beyond the biological warranty period is a consequence of the distorting effects of advances in biomedicine that are making their realisation more likely. In the following chapter, I provide a fuller argument in support of the value of longevity. In what follows, I explain what gives our lives meaning and value and the role of death in shaping the significance of our lives before considering two theories about the nature of modern technology.

Meaning, value and culture

Just as Kirkwood was influenced by Weismann's account of the germ and soma cells, so too was Sigmund Freud. In *Beyond the Pleasure Principle* (1920), Freud provides a description of the difference between the sex and death instincts, drawing on Weismann's distinction to provide a biological basis for them. The instincts are inherently conservative, seeking to restore an 'earlier state of things' (pp. 308, 331). The death instincts, for internal reasons, strive to return organic matter to its previous, inorganic state where *'the aim of all life is death'* (p. 311, emphasis original). By contrast, the sex instincts, which comprises the sexual instinct proper and the instinct for self-preservation, strive for self-renewal and the prolongation of life, evidence of which can be found in the 'coalescence' of germ cells, that is, reproduction (p. 316). The purpose of life, Freud maintains, is dualistic; life consists of the conflict and compromise between the emergence and continuation of life and the striving towards death (Freud, 1923, p. 381). Ambivalence, the awareness of this duality, is a 'fundamental phenomenon' that represents an incomplete fusion of the instincts (p. 382). Zygmunt Bauman (1992, p. 21) argues that Freud, rightly, leaves his readers to infer that this fusion can never be complete.

The ambivalence of life can lead many to view it as absurd. On reflection, the purpose of life appears to be to procreate and die, which the Disposable Soma Theory supports (although Kirkwood (2000, p. 16) rejects the fatalism this implies), with our lives lacking all significance, other than as a conduit for our genes. Our role in continuing the species offers little consolation: given the enormity of space and time, the human species possesses as much significance in the universe as an individual of the species. Without a meaningful reason for existence, the purpose of life becomes simply to die. It is this natural process of turning inorganic matter into organic matter so that it can return to inorganic matter that strikes some as absurd.[18]

In his assessment of Freud, Bauman (1992, p. 23) observes G. W. F. Hegel's claim that the history of culture is a record of how people respond to death. There are two obvious responses to the sense of absurdity: to avoid death or to confront it. It may be possible to avoid the sense of the absurdity of life by avoiding the issue of death. Over the last century, Geoffrey Gorer (1955, p. 51) argues, natural death (death that is the result of the ageing process) has become taboo and almost shocking, while violent death is an increasing and ever-present aspect of different fictional media. One explanation for this 'prudery' (as Gorer describes it) concerning natural death is that it is an attempt to hide our failure to gain control over nature. Bauman (1992, pp. 132–3) argues that the search for mastery is a defining feature of the modern age, where it provides freedom from necessity, from the laws of nature. In contrast to violent death, natural death is inevitable because it is intrinsic to human beings; it is a necessary feature of our nature.[19] The lack of control over when we will die and how long we can live, Bauman proposes, is shameful for the modern age because it represents a failure of mastery, and so natural death is hidden from view (p. 134). The idea of ageing as a disease rejects the inevitability of death because theories such as that of the Disposable Soma Theory and the potential of bio-technologies indicate that there can be some degree of control over death. Yet, controlling death does not involve confronting it but avoids death by giving the appearance of being able to abolish it.

To confront death, Martin Heidegger (1962, pp. 279–311) argues, is to have an authentic attitude towards it by recognising that we will die and cease to be.[20] In recognising this, we will realise that our lives matter to us; we will also realise that we are responsible for our lives because our lives are ours to live. To live a truly authentic life is con-sciously to choose how to live one's life; it is to provide it with meaning, and thereby make it purposeful. For Albert Camus (1955), avoiding the angst of the absurd involves three aspects: revolt against the meaning-lessness of existence; the freedom gained by this revolt to shape the meaning of one's life; and the passion to pursue one's own meaning. To confront death, therefore, is to realise and accept the freedom one has to provide one's life with meaning, leading Jean-Paul Sartre (1973, p. 28) to proclaim that we are what we make of ourselves.

To what extent people do and need to reflect upon their own death for them to appreciate that they provide their own lives with meaning and purpose is open to question. What is important to take from this response to death is the idea that it is the content of our lives that pro-vides them with meaning and in so doing makes them valuable. What matters in our lives are the activities and attachments that we maintain and pursue.[21] I use the phrase 'activities and attachments' to capture the whole range of features that make our lives what they are and provide them with their distinctive pattern. They are such features as our daily activities and fulfilling our basic needs and interests; our projects and goals; and relationships with family, friends, colleagues and other forms of social interaction. This is not to claim that people need consciously to pursue certain activities and attachments for their lives to be mean-ingful or to be distinctive; most people do not set about providing their lives with a unique narrative pattern.[22] Indeed, as Bernard Williams (1973b, p. 87; 1976b, p. 12) observes, it is testament to the success of their activities and attachments at providing life with meaning that, for many people, the meaning of their lives does not arise.

While Heidegger maintains that to have an authentic attitude to death we must decide how to live our lives, and ultimately choose the meaning of our lives, he also argues that our freedom to do so is constrained. Heidegger does not write of people, but of Dasein (literally, 'there-being'), which is thrown into the world and consequently finds itself in a particular environment with the fundamental needs and requirements of human beings. The issue of Being is of fundamental importance to Dasein in contrast to all other entities that exist because only Dasein is concerned with its Being.[23] Dasein's Being is expressed through the numerous ways in which it engages with other entities. This engagement itself reveals the Being of these entities. By describing people as Dasein, Heidegger can emphasise the fact that humans are thrown into human nature, which is a constraint on the choices people can make. Further, people are thrown into society: Dasein is fundamentally a Being-with-others, but also a Being in a particular cultural and social structure, which Dasein cannot have shaped, but which will shape and limit the choices they can make. Dasein is, therefore, thrown into what Heidegger (1971, p. 178) calls the clearing, a background of social and cultural practices. It is the clearing that provides the basis for Dasein's beliefs.[24]

The cultural background reveals the shared commitments of society concerning what activities and attachments are important and what makes life meaningful and valuable. An essential feature of the background is that much of the knowledge it provides cannot be articulated. Our cultural practices guide our activities and how we respond to each other, but this knowledge is a form of knowing-how rather than a knowing-that. Heidegger's account of the background has similarities to Ludwig Wittgenstein's notion (1967, sects 27–32) of the form of life and ostensive definition. It is by participating in the practices and institutions of the form of life that we learn about various social rules, for instance. Hubert Dreyfus (1993, p. 294) uses the example of physical distance in various social settings to emphasise this point. We are not taught explicitly how far to stand from family, friends or strangers, but cultures can exhibit differences in the physical distance that is expected in various social situations. This knowledge is acquired by experiencing and participating in a culture and is not gained through study.[25]

As described by Heidegger, the clearing will differ between cultures. Wittgenstein's concept of the form of life, however, is not simply a cultural and social phenomenon, but also embodies biological and natural facts associated with a form of life. Thus, while there are different cultures, they are united by facts about human beings (Heidegger's concept of Dasein is designed to emphasise these features). The three features of birth, relationships and death are some of the fundamental facts about human beings that are common throughout all cultures and on which different cultural responses and social institutions (Heidegger's clearing) are based.[26] This is not to suggest that the cultural response to these three fundamental facts will be the same throughout a particular culture, or that they are the only salient facts about people. As T. S. Eliot (1962) argues, sub-cultures are rooted within the culture of a whole society. It is the culture of a whole society that provides the clearing, the cultural background of shared social commitments.

Heidegger's account of the clearing indicates that we are not free to choose any meaning for our lives. We cannot step outside of the conceptual scheme that the clearing provides us with and which grounds our beliefs, concepts and values (Wittgenstein, 1967, sect. 103; Davidson, 1974). There are two important points to emphasise about this account of the cultural background. First, Dasein's view of the world and its understanding of Being will be shaped by the cultural practices into which it is born, but Dasein in its turn can alter these practices. The cultural background constrains our view of the world but it does not determine them. Second, the background of cultural practices does not imply that morality, for example, is fundamentally subjective and is 'projected' onto the 'fabric' of the world.[27] The cultural background consists of the cultural practices that are shaped by and grounded upon fundamental facts about people. It is these fundamental facts about human beings that give rise to our moral concerns and which are acknowledged, reflected and understood through the cultural and social practices of society. As a consequence, radically different cultures may appear to have different moral values, but at a basic level they reflect these fundamental moral concerns.

Technology and Being

With this account of the cultural background and what gives our lives value it is now possible to consider the relationship between technology and our values. I will compare and contrast the ideas of Hans Jonas and Heidegger. Both maintain that there is a substantive difference between pre-modern and modern technology, but while Jonas considers this difference to be one concerning the nature of technology, Heidegger claims it is a difference in our perception of the world.

Technologically determined objectives

For Jonas (1979), modern technology is restless in comparison to pre-modern technology. With pre-modern technologies, equilibrium would exist between the technological means and the ends to which they were applied for long periods of time. Human ends for which technology was required, such as constructing a bridge across a river in order to shorten trade routes, rarely changed, and when they did, they did so gradually. The technology needed to construct a bridge would develop until a saturation point was reached, whereby sufficient technological advancement had been reached to achieve the desired ends. Pre-modern technology was not experimental, theoretical or deliberately innovative.

In contrast, Jonas argues, modern technology is restless because it is in a continuous state of innovation. The development of technology to achieve our objectives will involve technological advances that suggest new objectives, which in turn require further technological development for them to be achieved (pp. 35–7). Unlike pre-modern technology, this process will never reach a saturation point because of the proposal of new ends. Jonas uses the examples of disposable plastic coffee cups, open heart surgery, artificial insemination and the defoliation of a forest by helicopter as objectives deriving from technological advances (p. 35). Knowledge of technological developments spreads more widely and quickly in the modern age than in pre-modern times because of industrial competition, which also drives technological innovation. Industries seek more efficient methods of developing their products, methods that will suggest new products and which must also be developed or adopted by other producers for them to remain competitive (p. 35).

Jonas fears that technology will eventually compromise humanity by changing its environment, values and the very biology of human beings (pp. 42–3). He points to advances in medicine leading to population growth as a threat to society and emphasises the implications for the environment of technological progress. Jonas dismisses the idea that technology can solve such problems because these solutions will lead to further problems (p. 36). Nuclear power, for example, will resolve the problem of coal shortages, but in turn raises new difficulties and concerns relating to the safe use of nuclear reactors and the storage of nuclear waste. For Jonas, rapacious technological progress threatens our very existence because human beings have become the subjects of technology and not its governors.

Jonas's account of modern technology raises a number of pertinent issues about what drives technological progress and the consequences of this. Technological advances have changed the nature of our objectives, such as the use of automobiles, trains and aeroplanes instead of the horse and cart, which do, as Jonas observes (p. 39), have a more intrusive and occasionally destructive effect on our world. His account of the proliferation of new objectives, driven by technological progress, also points to concerns about unabated consumerism. Nonetheless, his criticisms are of only limited effect. In the case of transport systems, technology has not changed the goal of getting from one place to another but has diversified the means for so doing. The fact that technology proposes new objectives need not, of itself, be problematic; they may be ones that we come to appreciate. Moreover, it is not the technology that proposes new objectives, but our imaginations and desires in response to technological developments. Should a 'technologically determined' objective prove detrimental to one or more of our values, it is because of the nature of and relationship between our values and not necessarily because they are a consequence of technological advances. This is in essence Heidegger's critique of technology, which I consider below.

A fundamental problem for Jonas's account is that he fails to substantiate a difference between pre-modern and modern technology. Heidegger (1967, pp. 272–3), for example, disputes the claim that pre-modern science and technology were neither theoretical nor innovative. Technological progress may have increased in recent centuries but this does not indicate a substantive alteration in the nature of technology. It might only demonstrate the effects of economies of scale in knowledge and innovation leading to an exponential increase in technological progress. Increases in life spans provide an ideal counter-example to Jonas's arguments. Whether increases in longevity are intentional or incidental, the search for longer life spans and the continuation of good health are not new objectives arising from technological progress. Rather, technological progress is improving the possibilities for realising these values. The issues that Jonas addresses do not concern an intrinsic feature of modern technologies but people and our values. Nonetheless, his observations indicate that we should question more the impact and consequences of the pursuit of our objectives.

A technological understanding of Being

Heidegger also distinguishes between pre-modern and modern technology but the difference is not one concerning the nature of technology, but how we understand the Being of entities. Dasein expresses its Being through its engagement with other entities, which in turn reveals the Being of these entities. It is through its engagement with entities that Dasein develops an understanding of Being. Dasein engages with entities in a number of ways, be it the use of a tool or to contemplate an object.[28] For Heidegger, the use of objects provides our primary understanding of Being and involves two aspects. It is on the one hand destructive, and so the creation of a pot involves distorting the earth to obtain clay; while on the other hand, it is also responsive to the Being of the objects that we use, revealing the nature of clay and bringing forth the pot (Heidegger, 1977, pp. 313–18; the example is from Alderman, 1978, p. 45). Harold Alderman observes that for Heidegger, 'Being is the ultimate "cause" of beings' (1978, p. 44). To be responsive to an entity is to engage with it in a way that is in harmony with its Being. In the modern age, this responsiveness is neglected and our interaction with entities is only destructive or aggressive. To demonstrate this relationship, Heidegger (1977, pp. 320–1) contrasts a windmill with a hydro-electric power station on the Rhine. The windmill is destructive because it is a construction made from natural resources and distorts the landscape, but it is also responsive because it emphasises the strength and freedom of the wind. A hydro-electric power station is only destructive because it absorbs and redirects the Rhine.

The modern age neglects the responsive aspect of our interaction with other entities because it involves a technological understanding of Being. Heidegger's concern is not with technology as equipment but drawing on Classical notions of *technē*, as a form of engagement with the world. The modern age is one dominated by reason and a rationalistic approach to the world. It attempts to capture the world in a mathematical picture, which only recognises those entities that can be depicted and encountered with certainty (Alderman, 1978, p. 36).[29]

All cultures attempt some degree of control over, and seek to find order in, their natural and social settings to some extent. Only in the modern age is there the view that complete control and order is possible (Dreyfus, 1993, p. 302). It is for this reason that the responsive aspect of engagement with other beings is neglected. Modern science, with its technological understanding of Being, gives the impression that the world can be ordered and controlled because it is rationally understood. As a consequence, science makes entities 'stand in reserve', waiting to be ordered again (Heidegger, 1977, pp. 322–3). Even human beings will be ordered and enhanced, that is, Dreyfus (1993, pp. 301–11) argues, made more efficient and flexible, where this process of ordering serves no purpose other than to demonstrate the ordering and controlling powers of modern culture. As a result, human beings are seen to dominate nature rather than be a part of it, where the entities of the world become resources for humans to use so as to make life more ordered and controlled.

Heidegger's account of technology is not simply an assessment of our impact on the environment, as the examples of the windmill and the hydro-electric power station might imply. His concern with technology is more fundamental than this. The cultural background shapes the way Dasein views the world; it provides Dasein with their understanding of Being, not just the Being of other beings but the Being of Dasein itself. As I understand Heidegger's account of technology, a technological understanding of Being threatens the pursuit of the good life. The background consists of shared practices that identify what activities and attachments are worth maintaining and pursing in order to give life its meaning and value. It is important, Heidegger argues, for the background to be made explicit to some extent so that people can know what is important in their lives. Nonetheless, the cultural background resists being made explicit because many of our practices cannot be articulated. In the modern age, with its rationalistic approach, what cannot be articulated is neglected as irrelevant, concealed or destroyed, just as Bauman claims natural death, as a necessity, is hidden from view.[30]

The attempt to articulate the cultural background leads to a loss of shared commitment to those things that make life fulfilling (Dreyfus, 1993). To articulate all aspects of the background undermines its grip on us and lays open our values to critical reflection and rejection. As a consequence, people become free, in the way implied by Sartre, to make of themselves what they will, to choose whichever activities and attachments they want to pursue but without appreciating what makes life good.

Heidegger does not maintain that the technological understanding of Being is the only understanding of Being in the modern age. If it were, he would be unable to criticise it as he does. Indeed, Dreyfus (1993, p. 307) proposes that the technological understanding might primarily be a Western phenomenon. While that may be so, Heidegger's ability to criticise it from within the Western clearing (assuming that there is a distinctive and common set of values that ground all Western cultures) implies that it is more appropriately associated with a sub-culture, albeit perhaps a dominant one. Heidegger's argument concerns the consequences of the technological understanding of Being supplanting all other understandings.[31]

One difficulty with Heidegger's argument concerns the responsive nature of our use of objects in the world. He maintains that the windmill is responsive to the world but that the hydro-electric power station is not. Yet, we might interpret the power station as emphasising the flowing nature of the Rhine and its power, and in so doing as being responsive to the Rhine's nature. The difficulty with Heidegger's account is appreciating how the use of an object can be in harmony with its Being because there are many ways of interpreting what constitutes a harmonious relationship. The cultural background identifies those values that contribute to the pursuit of the good life, and so one way of appreciating whether the use an object is in harmony with its Being is whether it promotes, or at least is not detrimental, to living well.

A second problem is Heidegger's support for a distinction between pre-modern and modern technology. Heidegger traces the origins of the rational and calculative view of the world associated with the technological understanding of Being to Socrates and Plato (Dreyfus, 1993, pp. 293–4). Plato insists on a method of deliberative measurement for comparing our values, one that would have the effect of minimising or eradicating the influence of the contingent aspects of our lives on the pursuit of the good life. In contrast, Aristotle recognises the role of the fragility of our lives in shaping our values (Nussbaum, 1986). For instance, our relationships and death are fundamental facts about human beings that ground our values, but they also represent vulnerable features of our lives: our interaction with others is subject to the vicissitudes of love, for example; and we will die, but when is uncertain. The debate between Plato and Aristotle is not necessarily as polarised as this suggests. Aristotle accepts that people do attempt to control the contingent features of life, through medicine, for example. Plato also recognises that to be free of the contingent nature of love would mean the good life would be devoid of something essential (pp. 165–235). What it points to, however, is a dispute about how much control there should be over our lives, such as over those aspects of our culture that we cannot articulate or that make the good life vulnerable, like a temporary state of depression or the length of life.

Heidegger's account of the technological understanding of Being is but a further contribution to the debate about how much we should seek to minimise the fragility of the good life. He does not identify a difference between the pre-modern and modern age because Plato's support for a deliberative method is indicative of a technological understanding of Being. What Heidegger's analysis of technology does provide is a greater appreciation of the diverse ways in which our lives resist being controlled. This is an essential feature of the clearing, which identifies the activities and attachments that are valuable and make life meaningful, and in so doing enable us to live well.

What remains to be considered is whether increasing life spans is indicative of a technological understanding of Being. To increase life expectancy beyond the biological warranty period will involve greater control over an importantly vulnerable aspect of our lives. Yet to do so risks threatening the cultural clearing because death, the uncertainty of when it will occur and the length of life are such pivotal features about human beings that they contribute, whether obviously or not, significantly to our values. Nevertheless, we also place great value on good health, and to maintain and restore health will involve prolonging life. Earlier, I proposed that prolonging life is not simply an incidental consequence of medicine but valuable as an aspect of the quality of life. If this is so, to increase longevity may not represent a technological understanding of Being but a harmonious activity that promotes the good life. It is to the value of longevity that I now turn.

2
The Misfortune of Death

To increase life spans is to delay death, and by exploring why death is bad for the one who dies we can appreciate the value of longevity. The idea that death is bad, however, has been questioned since ancient times. Epicurus and Lucretius raise substantive doubts about how death can be a misfortune for the deceased and in so doing question the value of living longer. In disputing these doubts and arguing that death can be a misfortune, my aim will be to show that longevity is an important value of the good life. Nevertheless, I will also argue that immortality would be undesirable and that it is possible to live too long, even when life appears to be good.

Before any discussion about the badness of death can begin it is important to have a definition of death. The *Oxford English Dictionary* defines death as: 'the end of life; the final cessation of the vital functions of an animal or plant' (I.1). In his assessment of the badness of death, Thomas Nagel (1970, p. 1) defines death as the end of existence, which might at first appear to conflict with the *OED*'s account. In many cases, if not most, the cessation of a human being's vital functions does not lead to the immediate end of their existence because their body remains present. This is so, however, only if we make a metaphysical assumption about personal identity, namely, that one is one's body, whether we are alive or not. Accounts of personal identity that specify the living body or psychological continuity as the criterion for identity will not find that the *OED*'s and Nagel's definitions conflict. On these accounts of personal identity, we will cease to exist when either one's biological vital functions cease or when one becomes permanently unconscious.[1]

Death as the end of one's existence is fundamental to the Epicurean and Lucretian objections to the misfortune of death, and it is this definition that I will adopt. Nevertheless, there are occasions where it will conflict with the account of personal identity that I accept and which will be relevant later in my discussion. It is beyond the scope of my discussion to provide a detailed account of personal identity, and consequently I will briefly state what I intend by it.

There are, I propose, at least three necessary aspects to our identities: psychological continuity, bodily continuity and our community. Psychological continuity, maintained by overlapping chains of connectedness and consisting of our memories, beliefs, desires and so forth, is important because it provides us with a sense of self. These elements shape and are shaped by our activities and attachments, which provides life with its distinctive pattern. We are, however, not pure psychologies: we are embodied beings. Our bodies shape and contextualise the possibilities that are open to us, as a species and as individuals. They are our mode of existing in the world; it is through our bodies that we interact with the world, and with other people. Indeed, many of the constitutive elements of psychological continuity, particularly memories, are dependent upon our physical perspective on the world. In some respects bodily continuity, constituted by spatio-temporal connectedness, is more crucial to questions of identity than psychology, as will become clear as my argument progresses.[2] A third aspect of our identities is the fact that we are members of a community (MacIntyre, 1985, pp. 33–4). Our relationships with family, friends, acquaintances and interactions with social institutions, shape, reflect and confirm our identities. Our relationships, such as being parents, children, spouses, lovers, employers and employees, all influence and reflect how we perceive and relate to each other.

One of the important outcomes of Derek Parfit's theory (1984) of personal identity is his argument that identity can be indeterminate, where it will not be clear that an individual is the same person. An example, for my account of personal identity, is a patient in a persistent vegetative state. With the irreversible loss of consciousness, an important criterion of identity is missing. Nonetheless, dispute about whether the patient continues to be the same person arises because their body remains, as does their communal identity: they will be recognised by their friends and family, and crucial to this recognition is the patient's body. Whether the patient survives whatever event led to their persistent vegetative state is questionable, which I discuss later in the chapter. What is presently important to emphasise is that these three aspects of identity are intimately bound, and an adequate account of them will involve providing a description of a person's biography.[3]

It follows from this account of personal identity that death is not always accompanied by the end of one's existence. Other things being equal, one's body will remain intact after its vital functions cease and one's communal identity will continue, albeit in a diminished form because it usually involves mutual interaction. At what point a person ceases to exist is unclear on this account, although there will be a point at which we do cease to exist. This need not be a problem for my argument. I will assume, in order to aid the progress of my discussion that death leads instantly to the end of existence.

Longevity and the misfortune of death

There are a number of misguided reasons for regarding death as bad for the one who dies. Death cannot be bad because of what will happen when we are dead. Death is the end of an individual's existence, and as such it is not a state of affairs that possesses any features that will affect us because it is not a state that possesses any features. Indeed, for this reason, it is doubtful whether we should refer to death as a state. The absence of any features may be what makes some regard death as bad. Nagel (1970, p. 3) observes that the attempt to imagine what it is like to be dead is commonly suggested as providing grounds for the fear of death. The problem with both the views that death is bad because it lacks any features and the fear of death on these grounds is that they are logically mistaken (p. 3). As the end of existence, it is impossible for there to be anything that it is like to be dead because there is nothing of which to conceive.

If it is impossible to imagine what it is like to be dead, then it is not obvious what can be known of death in order to assess it as being bad for one. It might be tempting to compare sleep with death. A particularly deep sleep offers some sense of what the absence of conscious experience entails but, unlike death, sleep offers only a gap in conscious sensation and so cannot provide an insight into the permanent absence of consciousness (Johnstone, 1976). Moreover, if a deep sleep were to be the complete absence of consciousness, albeit temporarily, it remains logically impossible for one to know what it is like to be asleep, and thus what it is like to be dead. The best insight that we can have of death is the death of other people, particularly of loved ones.[4] The death of those with whom we have personal relationships and their absence from our lives can make us reflect upon what makes our lives fulfilling and so what is lost when we die.

To reflect upon what is lost when we die will also reveal why death is a misfortune. Death is bad, Nagel (1970) argues, not because of any positive features but because of what it removes. Death is bad because it brings life to an end and in so doing deprives us of the 'goods that life contains' (pp. 1–2). As with most goods, other things being equal, more of them will be better than less.[5] Nagel does not describe these goods, but suggests some of them are sufficiently general as to be constitutive of human life, such as desire and perception (p. 2). If the sum of negative experiences outweighs the sum of positive experiences, the fact of experiencing 'is emphatically positive', which always makes it good to be alive (p. 2). For Richard Wollheim (1984, pp. 267–9), phenomenology is of such potency that it is difficult for us relinquish our 'longing' for it. Death will always be bad according to this account because it deprives individuals of experiencing, a view Nagel (1986, p. 225) supports. This does not imply that someone might prefer to continue living a life of predominantly negative experiences than to die. In such a situation, continued life is worse than death, where both options are bad. Wollheim (1984, p. 267) describes the choice as one between a shorter and a longer life, rather than between life and death. Nagel also notes that this account of the misfortune of death maintains that the value of life derives from its content and 'does not attach to mere organic life' (1970, p. 2). This is not to deny that our bodies are without value, but they are valuable only because they are necessary to support the content of our lives.

Williams (1973b) concurs with Nagel's view that death is bad for the one who dies because it deprives them of the future goods of life. Williams describes these goods as the *praemia vitae*, the 'the rewards and delights of life', which our activities and attachments provide (pp. 83–4). He goes further than Nagel and grounds the possession of the *praemia vitae* on a person's desires. Many desires that people have are conditional upon their existence because, 'many of the things I want, I want only on the assumption that I am going to be alive' (p. 85). Yet, Williams claims that there are some desires, categorical desires, which are not contingent upon our existence. He uses the example of a forward-looking rational suicide who ponders what possibilities lay before him or her and considers whether they want to actualise them. If the rational suicide decides to actualise a possibility, they do so because of a categorical desire (pp. 85–6). It is the suicide's desire to fulfil the possibility that propels them into the future; the desire to fulfil this possibility provides the suicide with a reason for living, at least for the duration of the fulfilment of the desire, and this desire is not conditional on their existence. We do not need to be potential suicides to have categorical desires. An alternative example is the desire that one's children be successful in their endeavours. Throughout life, a person might work to ensure that this desire is fulfilled, but the desire is not conditional on them being alive. Death would prevent this person from enjoying their children's successes, but the desire can still be fulfilled, and they would want it to be fulfilled even when they no longer existed.

Categorical desires are more than a simple 'reactive' drive for self-preservation. Williams recognises such a desire, which, although necessary for continued existence, is insufficient on its own for this purpose (p. 86). People have many categorical desires, which need not be 'very evident to consciousness, let alone grand or large', but they are what make a person keep on living (Williams, 1973b, pp. 84–5; 1976a, pp. 10–12, quote on p. 12). They are the product of our dispositions, needs and interests and give rise to the activities and attachments that we maintain and pursue and which provide life with its meaning and value.[6]

One difficulty with Williams's account is that it does not appear to be capable of explaining why, if at all, the death of a foetus or infant is a misfortune. It is doubtful that an infant can have such sufficiently developed categorical desires for their death to constitute a misfortune for them (McMahan, 2002, p. 182). Nevertheless, Williams's account (1973b, p. 84) of the goodness of the *praemia vitae* provides a response to this problem. If possessing the *praemia vitae* is good and possessing the *praemia vitae* for longer is good, getting to a point where one possesses them must also be good. The death of an infant is, other things being equal, a misfortune for them because they will not have got to a point in their life where they possess the *praemia vitae*, or indeed, have sufficiently developed categorical desires to provide them.

Nagel and Williams agree that if life is good, to have more of it is better than to have less of it. Death is a misfortune because it results in less of a good life. What distinguishes Williams's account of the misfortune of death from Nagel's is what grounds this misfortune.[7] For Nagel's view, experience, independent of its content, is a good and its loss is a misfortune. On Williams's account, the misfortune of death is dependent upon the content of our experiences, specifically those experiences that derive from our categorical desires and which ground our activities and attachments. If life is to be meaningful and valuable it is because of the content of our experiences, and not the fact of simply having experiences. It follows from Williams's argument that death need not always be a misfortune for the deceased because there may be situations where their activities and attachments fail to make their lives sufficiently fulfilling for them to want to continue living, and nor will they in the future.

The reason why death is, other things being equal, a misfortune for the one who dies, reveals the value of longevity. In the previous chapter, I noted that longevity is valuable as a structural feature of the good life. When life is good, we want more of it: more life is better than less. Endeavours that can increase life spans while also maintaining the quality of our health are valuable because they will enable us to continue experiencing the fulfilment of our categorical desires. This implies that increasing life spans does not represent a technological understanding of Being because the pursuit of the good life is a fundamental value for human beings. As a value that is central to the good life, to increase longevity is in harmony with its pursuit, for the simple reason that if our lives are good, then it is also good that they continue.

Epicurean and Lucretian objections

Williams and Nagel provide accounts of the badness of death grounded upon the fact that death brings a fulfilling life to an end. Nonetheless, any theory that describes the badness of death in terms of the goods of which it deprives the deceased must resolve a number of problems put forward by Epicurus and Lucretius. On the basis of these problems, Epicurus and Lucretius deny that death is bad, and consequently that a longer life is desirable.

The first problem proposed concerns the phenomenology of the badness of death. Epicurus argues that 'death is of no concern to us, since all good and evil lie in sensation and sensation ends with death' (1964, p. 54). Even if we reject Epicurus' hedonism, his argument still poses the problem that as non-existence death cannot involve the awareness of experiences, be they misfortunes or otherwise. Death may be a misfortune according to the deprivation thesis, but it is not one of which the deceased can be aware. This fact about death implies that it cannot be a misfortune for the deceased because only those things we consciously experience can be good or bad for us.

A second Epicurean problem is of an ontological nature.[8] Epicurus maintains that death is the end of one's existence and as such cannot be bad for the deceased, 'for while we exist death is not present, and when death is present we no longer exist' (p. 54). This creates two difficulties for the badness of death, concerning the lack of a subject, which in turn entails the problem of timing. If death is the end of existence, there is no subject for whom death is bad, and as the subject no longer exists, there can be no time when they experience this misfortune, raising the question of when and for how long this misfortune endures.

Lucretius (1994, III, 871–9) raises a third problem for the claim that death is bad for the deceased concerning an asymmetry in our attitudes towards non-existence. He points out that just as in death we do not exist, so we do not exist before we are born. If death is a misfortune because it involves our non-existence, so our non-existence prior to birth must also be a misfortune. Yet, we do not normally regard pre-natal non-existence in this way, which gives rise to an asymmetry in our attitudes to non-existence that is difficult to justify. Even when we consider why the non-existence of death is a misfortune, because it deprives us of the goods of life, the asymmetry remains: had we been born earlier than we actually were, we could have benefited from more of the goods of life. If we do not view pre-natal non-existence as a misfortune, we should not, on the Lucretian argument, regard death as a misfortune.

If the Epicurean and Lucretian objections prove successful and death is not bad for the deceased, the value of increasing life spans becomes questionable. Their arguments that 'death is of no concern to us', if successful, mean it can never be bad to die, no matter what age this occurs. It would not be possible to die prematurely because this implies that death is a misfortune (Luper-Foy, 1987, pp. 238–9). Lucretius adds that it cannot matter when we die because we will all be dead eternally, and so '[t]he period of not-being will be no less for him who made an end of life with today's daylight than for him who perished many a moon and many a year before' (1994, III, 1087–94). This is so because people who die at different ages and at different times will each be dead for an infinite duration, which because of the nature of infinity, will be the same for each person.[9] We should accept, Lucretius argues, that we have a 'fixed term' and instead turn our attention to contemplating the nature of the universe (III, 1075–84). To accept that death is not a misfortune will take away the desire for immortality, which in turn will lead to a happier life (Epicurus, 1964, p. 54).[10] Lucretius (1994, III, 1080) questions what the 'lust for life' is that makes people fear 'uncertainties and dangers', which echoes the earlier critique of the technological understanding of Being. To pursue longer life spans might reflect this understanding, and rather than helping us to flourish, in fact undermine it.[11]

The phenomenological problem of death

The phenomenological problem is the claim that death cannot be bad for the deceased because they cannot experience this badness. This problem rests on the assumption that for anything to be good or bad for a person, they must be aware of it. Nagel (1970, pp. 5–6) provides two counterexamples to demonstrate that this is not the case.

In his first example, Nagel argues that an act of betrayal is bad for the betrayed person, even if they never come to discover that they were betrayed. The nature of betrayal is important for this counterexample. Betrayal is not bad because it makes the betrayed person suffer. Rather, the betrayed person suffers because betrayal is bad. It is this feature, the badness of betrayal, which implies that when a person is betrayed they suffer a misfortune, whether they are aware of it or not.

Nagel's second example is that of an intelligent adult who, after a severe accident, is reduced to the mental condition of a contented infant. The adult does not object to their new state because they are not aware that they were once an intelligent adult. Moreover, it is not the case, other things being equal, that having the mental condition of a contented infant is a misfortune. Other things, however, are not equal, and it is not an infant who has this condition but a formerly intelligent adult who is reduced to this state. Only a consideration of the state of the person before the severe accident, and what their life would have been like had the accident not occurred, reveals that the intelligent adult has suffered a misfortune, but it is not one of which the formerly intelligent adult can be aware.

Nagel proposes that what these counterexamples to the Epicurean argument demonstrates is that a person's life 'includes much that does not take place within the boundary of his body and his mind, and what happens to him can include much that does not take place within the boundaries of his life' (1970, p. 6). It is, Nagel argues, arbitrary to restrict what is good and bad only to the intrinsic properties of life. Our lives consist of both the subjective and objective points of view. One's experience of life is obviously subjective, but it is influenced by the objective point of view. It is the objective aspect of our lives that provides the social dimension of our identities. And the objective consideration of an individual's life reveals the relations that they have with other people and the world in general. We can, to some extent, consider our own lives from this objective perspective: if we could not, then the sense of the absurdity of life might never arise. Nonetheless, we can never occupy a fully objective point of view on our lives because we can never reject or step outside of the subjective view, as Nagel's counterexamples demonstrate.[12]

The misfortune of the formerly intelligent adult can be recognised by examining the history of the adult (that they were an intelligent adult), what they are now (an adult with the mental condition of a contented infant), and what they would be (an intelligent adult) were they not to have had the severe accident. Such an assessment of the intelligent adult's life can only be made by viewing their life objectively and considering the nearest possible world in which the severe accident did not occur. The misfortune that befalls the intelligent adult is not, from a subjective perspective, a state of affairs the intelligent adult can experience. Indeed, from a subjective perspective, the formerly intelligent adult is quite content with the mental condition of an infant. From an objective perspective, when considering the nearest possible world in which the intelligent adult did not have the severe accident, it is clear what misfortune has befallen them. It is not a misfortune that the intelligent adult can ever be aware of, but it is their misfortune. Such is the case with the badness of death. Unlike the intelligent adult, however, there is some doubt about whether someone exists for whom their death can be a misfortune.[13]

The ontological problem

The ontological problem maintains that death cannot be bad for the deceased because there is no subject for whom death can be bad. This difficulty for the badness of death involves two issues: the non-existent, and therefore missing subject, and when the deceased experiences the misfortune of their death.

In some respects, the first problem, that of the missing subject, is relatively straightforward to resolve. The Epicurean claim is that the statement, 'death is bad for X' makes no sense because there is no X for whom death can be bad. The statement referentially fails and so is false or does not make sense. What matters, however, is that X once existed. There may be some dispute about how the statement 'death is bad for X' can refer, where, for example, we could adopt Bertrand Russell's idea (1905) of providing a definite description for X; or John Searle's family (1958) of descriptions for X; Peter Strawson's claim (1950) that the use of proper nouns is context-dependent; or Saul Kripke's argument (1981) that a proper name is a rigid designator. What matters is that X once existed. If this is the case, there is someone to whom the statement 'death is bad for X' can refer, and so there is someone for whom their death can be a misfortune. What is more difficult is ascertaining when death is a misfortune for the one who dies.

When is death bad?

Death is a misfortune because it is the end of existence. The fact that the deceased no longer exists, however, entails there can be no time during their existence when they suffer the misfortune of death. Even taking into account the claim that we can suffer a misfortune without being aware of it, it is less obvious that we can suffer a misfortune when we do not exist.

One solution to the problem of timing is to argue that the badness of death is experienced during one's life. Death will frustrate our experiences of the fulfilment of our categorical desires. This badness is a feature of our desires from the moment we acquire them, although we will be unaware of this. This is not, as Geoffrey Scarre (2007, p. 94) observes, to promote the idea of backward causation, whereby future events affect past events. Rather, it is to claim that our death will make it true of our frustrated desire, that in acquiring it, we have experienced something bad (Pitcher, 1984). For example, whether we have undertaken sufficient work to pass an examination will depend largely on our success in the exam: if we fail, other things being equal, then we did not do enough work (Scarre, 2007, p. 94). Such a solution to the problem of timing seems counter-intuitive, not least because it conflicts with the proposed explanation for the misfortune of death. Death is bad, when it is bad, because of the future goods of which it deprives the deceased. The badness of death derives from the absence of these future experiences and not our present experiences. Moreover, the argument would seem to imply that, when our death is bad, much of our life has also been bad for us because we acquired desires that will be frustrated. It does not necessarily follow that if an event is bad for us in the future it must be bad for us now. We might fail an exam, but benefit greatly from the work we undertook before taking it (pp. 93–8).

Nagel's (1970, p. 6) solution to the timing problem is to argue that the misfortune of death is atemporal, that is, it lacks a location in time. He draws on his argument that to resolve the phenomenological problem of death it is necessary to regard the deceased's life from an objective point of view. There are two noteworthy difficulties with his solution. First, Ben Bradley (2004, p. 2) argues that in proposing that the badness of death is atemporal, Nagel requires it to be a misfortune like no other. It might appear that death is unlike other misfortunes because it involves the problem of the missing subject. The example of a persistent vegetative state suggests otherwise, although the role of the body in questions of personal identity complicates matters. It may be that the existence of a living body or a corpse is sufficient to identify the subject, in which case death, for a time, does not involve the problem of a missing subject. If the irreversible loss of consciousness is sufficient for the end of one's existence, the problem of a missing subject also affects those in a persistent vegetative state.[14] In either case, it would appear that Bradley's observation is correct.

Second, an atemporal account of the misfortune of death cannot explain whether death is bad for the deceased. Nagel (p. 3) points to an asymmetry between life and death. The goods of life, he argues, can be attributed to an individual at each point of their life. This is not the case with death because it consists of nothing: what is bad about death is the loss of the good things in life and not the presence of bad features. Thus, J. S. Bach had more of the goods of life than Franz Schubert because he lived longer, but death 'is not an evil of which Shakespeare has so far received a larger portion than Proust' (p. 3). William Shakespeare has been dead for longer than Marcel Proust but because death lacks attributes, he cannot have acquired more of the misfortune of death than Proust. It is because death lacks attributes, along with the fact the deceased no longer exists, that Nagel maintains the misfortune of death occurs atemporally.[15]

Situations can arise, however, where death is not a misfortune for the deceased, but to explain whether death is bad or not requires that the misfortune of death is temporal. For example, Christopher Belshaw (2009, pp. 76–84, 94–127), among others, argues that in order to assess whether death is bad, it is necessary to consider the life the deceased would have lived had they not died when they did.[16] Thus, if X dies at t but they would have received a job promotion at $t1$ had they not died, then death has deprived X of some good. Were X to have been in a car accident at $t2$ and have been reduced to the mental condition of an infant, death would have deprived X of an event that they might conceive of as being worse than death. Only by taking into account the goods that death deprives the deceased of is it possible to comprehend whether death is bad for them.

The difficulty with a temporal explanation for when a person suffers the misfortune of death, and what motivates Nagel's atemporal response, is the lack of a subject. The difference between real changes and so-called 'Cambridge' changes can provide a solution (Scarre, 2007, pp. 105–10; Grey, 1999, p. 361). A real change involves an alteration in the intrinsic properties of a thing at a particular time; Cambridge changes are relational changes arising as a consequence of real changes. For example, when a spouse dies, they undergo the real change of ceasing to exist; their partner undergoes the Cambridge change of becoming a widow or widower. The partner does not undergo a change to their intrinsic properties but does so in their relational properties (Scarre, 2007, pp. 105–6).

For a person's death to be a misfortune for them, it must deprive them of goods, such as a job promotion, where the fact that certain real changes did not occur creates relational changes that affect the subject. Scarre maintains that not all relational properties that might affect the misfortune of death relate to changes in intrinsic properties (p. 108). A person might die before they reach the age of thirty, which for them would have been a defining age in their lives. In this case, the relationship is with time. Whatever the nature of these relational properties, although most will be with intrinsic properties, they ground the misfortune of death in time. Furthermore, the relational aspects of life only become fully apparent when we view our own and other's lives from an objective point of view.

Determining the misfortune of death

A temporal account of death risks implying that Shakespeare's death is a greater misfortune than Proust's because he has been dead for longer than Proust and so has been deprived of more of the goods of life, assuming their lives would have been good (Belshaw, 2009, pp. 80–1). Indeed, Nagel (1970, p. 8) argues that death deprives people of countless future possibilities. To avoid this conclusion, it is necessary to consider how long they could reasonably have expected to live had they not died when they did. Given the average life expectancy of approximately 80 years, both were deprived of roughly the same amount of life (Belshaw, 2009, p. 81).[17] In addition, it will also be necessary to consider whether the years of which their deaths deprived them would have been on the whole, good. If this were the case, then Shakespeare and Proust have suffered similar misfortunes. The assumption about how long a person could expect to live raises the further problem of 'over-determining' the misfortune of a person's death.

The problem of over-determination becomes most apparent when comparing the deaths of individuals early in life and later in life. For example, John Keats's death at the age of 25 is generally regarded as being a greater misfortune than Leo Tolstoy's death at the age of 82.[18] Keats has been deprived of considerably more years of life than Tolstoy hence Keats's death is the greater misfortune, assuming these years would have been good. Jeff McMahan (1988) observes that the deprivation thesis unquestioningly assumes the truth of the antecedent of the conditional, 'if X had not died'. It is not always the case that in the nearest possible world the antecedent is true. The nearest possible world is one that most closely resembles the actual world but where the immediate causal events of X's death are missing (McMahan, 1988, pp. 45–9).[19] Take Joe, also aged 25, who is killed by a bus while crossing a road on his way to a medical appointment.[20] The deprivation thesis maintains that his death is a misfortune (assuming the consequent of the conditional would have been, on the whole good) and a greater misfortune than Tolstoy's death. Yet, in the nearest possible world where the antecedent, 'if X had not died', is true, Joe might have discovered during his medical appointment that he had cancer and only three months left to live. When Joe was killed by the bus, he was deprived of only three months of life. Had Tolstoy not died when he did, he might have lived for more than three months. The deprivation thesis struggles to explain the general belief that Joe's death is a greater misfortune than Tolstoy's because Joe has been deprived of the same or fewer months of life than Tolstoy.

If we are to explain why Joe's and Keats's deaths are a greater misfortune than Tolstoy's, then we must point to the life that Joe and Keats have led in comparison to Tolstoy's life. This much is implied by Williams's claim that if the *praemia vitae* are good, then getting to a position where one possesses them must also be good. In dying at the age of 25, Keats and Joe have been unable to develop and benefit from their activities and attachments to the extent that Tolstoy has, and this is so in whichever possible world one considers where Keats and Joe die young. Indeed, Nagel's discussion (1970, p. 3) of the atemporal nature of death alludes to this when he observes that Bach has had more of the goods of life than Schubert. What this solution implies, however, is that the deprivation thesis alone cannot explain why death is a misfortune for the deceased.[21] This should not be surprising when we consider the earlier example of the misfortune suffered by the intelligent adult. It is only by taking account of the life of the intelligent adult before their accident and after it that we are able fully to appreciate their misfortune.

The normal life span

There are two points to emphasise about the issues thus far outlined. The first concerns the normal life span, which provides part of the background context against which assessments of the misfortune of death are made. The normal life span identifies the length of life for which an individual might reasonably expect to live. It is because of the normal life span that Shakespeare's death is not a greater misfortune than Proust's. As a reasonable expectation for people, it will play a central role in assessing that Keats's and Joe's deaths are greater misfortunes than Tolstoy's. Furthermore, because it is part of the background context, it will not vary between the actual world and the nearest possible world. What makes Joe's death a greater misfortune than Tolstoy's, despite the fact Tolstoy may be deprived of more of the goods of life, is the fact that Joe has not experienced as much of the fulfilment of his categorical desires as Tolstoy has of his. This claim makes sense, however, only against the context of the normal life span because it depends on the assumption that Joe could reasonably have expected to experience the fulfilment of his categorical desires for longer than he did. If there were not this belief about how long Joe could expect to experience the goods of life, then Shakespeare's death would be a greater misfortune than Proust's because he has been dead for longer.[22]

The normal life span does not refer to a statistical account of the average length of life, nor does it refer to scientific evidence for a biological limit to the human life span, such as the Disposable Soma Theory. Rather, it depends upon beliefs and traditions, like the biblical claim that the human life span is 70 years, which are supported by observations that people do not generally live beyond an approximate age. Just as the normal life span provides the context against which we assess the misfortune of death, so it also provides the context against which we choose our activities and attachments. The fact that we can reasonably expect to live for approximately 80 years contributes to the significance and value of our choices. For example, a project requiring ten years of commitment will, other things being equal, be more valuable than a project lasting a year because it involves committing a substantial part of our lives to it. Were the normal life span to be a million years, this would alter the value of our projects and goals, a point I return to at the end of this chapter.[23] This is not to claim that we consciously choose our activities with the knowledge that we have a restricted number of years in which to pursue them. As the context against which we make our choices, the normal life span is a feature of the cultural clearing that shapes our values and reveals which activities and attachments are worth maintaining and pursuing.

The subjective view

The second point concerns the subjective view on death. Tolstoy lived for as long as he could reasonably have expected to live, and in so doing was able to experience more of the fulfilments of his activities and attachments than Keats was able to experience of his. The comparisons of the misfortune of death between Keats and Tolstoy or between the counterfactual assessments of Joe's death necessarily take place from an objective perspective, but there also remains the perspective of the subjective view. Tolstoy might agree that Keats's death is a considerable misfortune for Keats, but nonetheless maintain that his own death at 82 is a considerable misfortune for himself.

In order to explain how death can be a misfortune for the deceased, it is necessary to consider an individual's life from an objective perspective, but in so doing it is important not to neglect the subjective perspective. Indeed, the subjective perspective provides the basis for the misfortune of death because death is bad for the one who dies when it deprives them of their experiences of the fulfilment of their categorical desires. In this respect, so long as an individual possesses categorical desires and they provide their life with fulfilling experiences, their death will always be bad for them. To appreciate the misfortune of death for the deceased requires reflecting upon the subjective view of a person's life, and the fact that from this view life is meaningful, valuable and worth continuing.

The subjective view on death means it is always a misfortune for the deceased, so long as their life is sufficiently meaningful to them for them to want to continue living. This will be so, no matter how old a person is when they die. At first, this might appear to contradict the idea of a normal life span providing the background context against which the assessment of the misfortune of death is made. But, as Nagel (1970, pp. 9–10) observes, the idea of a limit to the length of life is not a feature of our subjective point of view on our experiences. Although Tolstoy has lived for longer than the normal life span, this will not, other things being equal, diminish his desire to continue living so long as he finds his life valuable.

The asymmetry of non-existence

The final objection to the view that death is bad for the one who dies concerns the asymmetry in our attitudes to pre-natal and post-mortem non-existence. Death is a misfortune for the deceased because it deprives them of the goods of life, but to be born when we were might also be regarded as depriving us of these goods. Our lives are, as Lucretius observes, bounded by non-existence. If we do not view pre-natal non-existence to be a misfortune for us, we should not regard death as misfortune. There are two prominent responses to the asymmetry problem concerning a bias in our attitudes towards the future and personal identity.

A bias towards the future

One solution to the asymmetrical attitudes towards non-existence focuses on the alleged bias that we have towards future events in contrast to those of the past.[24] Parfit (1984, pp. 165–7) uses the following thought experiment to identify this bias. A patient must have a painful operation for which they will not be anaesthetised. Instead, they will be given a drug after the operation that will make them forget it. The patient awakes, not remembering falling asleep and wonders if the reason for this is because they have had their operation. A nurse is aware of the information for two patients but does not know which applies to the newly awakened patient. The patient either had their operation the previous day, in which case it lasted ten hours, or they are due to have their operation later that day and it will last one hour.

Parfit argues that most people, if they were in the same position as the patient, would prefer that their operation had taken place on the previous day, even though it lasted ten hours. The pain from the ten-hour procedure may have been considerable because of the duration of the operation, but it is in the past and, moreover, the patient cannot remember it. Although the one-hour operation is shorter, it is still to be endured and so the patient would have to experience an hour of pain. The fact that the patient must endure future pain is of more concern to them than their past pains. This thought experiment is intended to show that, in general, people are biased towards future states of affairs as opposed to those in the past. If this observation is correct, it is not irrational to be more concerned about post-mortem non-existence than pre-natal non-existence because the former is in the future while the latter is in the past.

A bias towards the future may explain why we are not irrational to have an asymmetrical attitude to non-existence but it does not show, however, that pre-natal non-existence is without misfortune. Nonetheless, the bias towards the future points to a difference between post-mortem and pre-natal non-existence, such that the former is a misfortune while the latter is not. Scarre (2007, p. 102) observes that because of the direction of time, and specifically the direction of causality from cause to effect, pre-natal non-existence does not prevent the fulfilment of our categorical desires, unlike death. Our ability to gain experiences from our categorical desires cannot extend into the period of pre-natal non-existence in the way that it can with post-mortem non-existence. We can undertake measures in order to prevent our death, thereby delaying the ensuing post-mortem non-existence. This cannot be done with pre-natal non-existence: an individual cannot endeavour to be born earlier than they actually were.

Being born earlier

A further objection, and one that lends support to the former argument, is the claim that were we to be born earlier than we actually were, we would not be the same person, that is, our identities would alter. On this basis, it makes no sense to be concerned about pre-natal non-existence. The direction of time and causal processes entail that the only way our categorical desires can extend into the past is if we had been born earlier, but if this were so, they would not be our categorical desires.

The objection relies on the idea that to be born earlier will involve a change of identity.[25] The reason for this can be explained by Parfit's *Time-Dependence Claim*: '[i]f any particular person had not been conceived within a month of the time when he was in fact conceived, he would in fact never have existed' (1984, p. 352). Had a person been born more than a month earlier, a different sperm and egg would be involved, leading to a genetically different person, where it is assumed genetic identity contributes significantly to our personal identity. For example, although Parfit argues that the criterion for personal identity is psychological continuity and connectedness, our psychologies will be shaped by our genes and environment, such that a different genetic identity will entail a different psychology. The Time-Dependence Claim, in allowing a person to be conceived within a month of when they actually were, admits that a person could derive from a different sperm and egg, and so would have a different genetic identity. I assume, for the present, that this is not possible, but will return to this issue later in the discussion.

With developments in modern biomedicine it is possible to freeze embryos, eggs and sperm, for future use. This means it is possible for a person to have been born at a different time (by more than a month) than they actually were while retaining their genetic identity. Whether or not being born earlier but with the same genetic identity will involve the same or a different person depends upon the relationship between our genetic identity and personal identity. For Kripke (1981, p. 113) what matters for a person to remain the same person over time is that they have the same origins, specifically, that they derive from the same sperm and egg. So long as Aristotle has the same parents, for example, he would remain the same person even if he never became a philosopher or taught Alexander (p. 62). On this basis, an individual could be born at a radically different time to when they actually were, while still retaining their identity, so long as their genetic identity remained the same.[26]

The Kripkean account of personal identity emphasises the importance of biology for identity, and in so doing allows for radical changes in biography (Kaufman, 1996, p. 308). A theory such as Parfit's, however, may involve a change of identity where there is a radical alteration to biography. If a person who was actually born in 1950 were to be born in 1850, and they somehow retained their biology, their biography would be so radically different they would have a different psychology. On the Parfitian account of personal identity, they would not be the same person.

On the account of personal identity that I provided earlier, the conclusion appears to be more complex. A person born earlier than they actually were would have the same biology, and hence the same body, but would have a different psychology and different communal relations. In many respects our bodies are crucial to questions of identity, but not necessarily as a way of resolving matters of dispute. For example, it is questionable whether the intelligent adult reduced to the mental condition of a contented infant is the same person after their accident (or that a person in a persistent vegetative state is the same person prior to their condition). The reason why the issue of their identity is indeterminate is because the intelligent adult retains their body after the accident. Indeed, it is because of this that friends and family are able to recognize their relationship to the formerly intelligent adult. The intelligent adult, reflecting upon what their life would be like should such a situation arise, is unlikely to regard themselves as having survived the accident. What is missing after the accident is the content of the intelligent adult's life. It is this content that provides the intelligent adult's life with its distinctive pattern and makes them who they are, and in so doing makes their life fulfilling. In the case of a radical change in psychology or the complete loss of psychology, bodily continuity and community identity are insufficient for a person to have survived whatever event brought about the change or loss of psychology.

If we were to be born at a significantly different time from when we actually were, but we were to retain our physical origins, the change in identity would be more radical than that of the intelligent adult. Not only would the content of our lives be different, but our communal identities would be different. As a result, the only way we could obtain the goods of life prior to the time of our births is to be born earlier, and this would involve a change of identity such that we cannot benefit from being born earlier. It is, therefore, not irrational to have more concern about our post-mortem non-existence than our pre-natal non-existence.

There is a problem with this solution. Personal identity is not an all-or-nothing concept. This much is admitted by the Time-Dependence Claim. It asserts that if a person conceived at time *t* were instead conceived at time $t1$, which is no more than a month later or earlier than t, they would remain the same person. The person conceived at $t1$ would be slightly different from the person born at t. They would have slightly altered genes and a slightly different psychology but significant factors would remain the same, such as their parents and their environment. As a consequence, these differences might not be sufficient for them to constitute two different people.

The fact that there can be an alteration in the time of conception without a change of identity means it is possible for someone to be born earlier than they actually were. How great the time gap between t and $t1$ would need to be for the person conceived at these times to become a different person is unclear. Nonetheless, the losses from death have the potential to be significantly greater than those from the time of our birth. This feature of death, when combined with a bias towards the future and the possibility of delaying death, might be sufficient to explain and justify the asymmetry in our attitudes towards non-existence.

The tedium of immortality

At this point, it will be beneficial to summarise the argument so far. My aim has been to show that death can be a misfortune for the deceased in order to demonstrate that longevity is valuable. Death is bad, when it is bad, because it deprives the deceased of the experiences they derive from the fulfilment of their categorical desires, which ground their activities and attachments. The discussion of the timing problem reveals that we cannot rely solely on the deprivation thesis to explain the misfortune of death. We must also take into account the life an individual has led, which can assist in explaining why dying earlier is, other things being equal, worse than dying later.

Epicurus and Lucretius pose a number of challenges for this conclusion. Nagel's counterexamples, and the need to acknowledge that our lives consist of both the subjective and objective points of view, mean we can explain why death is a misfortune for the deceased even though they cannot be aware of it. The fact that the deceased once existed resolves the problem of the missing subject, while an account of the relational changes that occur after the deceased's death provides a temporal solution to the timing problem. Our bias towards the future and the change to our identities if we were to be born earlier than we actually were suggests two interdependent solutions to the asymmetry problem, although they are by no means conclusive.

Death is a misfortune for the deceased because when life is good, more of it is better than less. Williams (1973b) oberves that this implies that immortality is desirable, but he argues that it is possible to live too long. In making this claim, he is not referring to a situation where an individual is in such unbearable pain, for example, that they consider it better to die than to continue living (p. 100). There are also many different types of circumstance whereby an individual is prevented from gaining fulfilling experiences from their categorical desires. In such cases, an individual's life may lack sufficient quality for them to want to continue living. Rather, Williams's concern is with a situation where an individual loses their categorical desires and becomes emotionally cold and bored with their activities and attachments.

As an example of a long life span, Williams uses the character of the 342 year-old Elina Makropulos from Karel Čapek's play, *The Makropulos Case* (1922). In his discussion of the 'tedium of immortality', as he refers to it, Williams makes two important caveats. The first assumes that every person in society will experience significantly increased longevity. In Čapek's play, only a few people lived as long as Makropulos. This avoids the added complication that people may become bored with life because they outlive their friends and family. Williams's concern (1973b, p. 90) is to emphasise the boredom of living. The second caveat maintains that despite significantly increased longevity, people remain at a particular age, 42 in the case of Williams's example. Thus, although Makroplulos was 342 years old when she died, she had remained 42 for 300 years.[27] Should life spans increase considerably, it is not obvious that people will remain at a particular stage of ageing throughout their lives. Nonetheless, increasing longevity by retarding the ageing process might lead to a disproportionately long 'middle age' and a healthier 'old age'.

To discuss the tedium of immortality, Williams considers increasing longevity through a series of lives (pp. 92–6). He proposes setting aside the issue of bodily continuity (which he considers necessary for personal identity) and presumes that some form of bodily continuity can be maintained between each life. Increased longevity, therefore, would consist of a series of births and deaths and rebirths with each successive life being bodily continuous in some way with the previous life. There are then, two scenarios to consider. The first is where each successive life will be different because each life will have a new character, including new memories. It is doubtful that any person would consider this to be a form of increased longevity. The concern to live longer is not simply a desire that some aspect of oneself will live on. This much is achieved through reproduction. The desire to live longer is the concern for the survival of oneself into the future. As argued, survival depends upon one recognising the content of a life as being one's own, as consisting of one's categorical desires, which provide the basis for one's activities and attachments. A person would not identify themselves with a future person whose character was radically different from their own, even if there was bodily continuity.[28] The alternative scenario would be where these series of lives could maintain psychological continuity. Each life would repeat experiences of the previous life as it matured, gained an education, fell in and out of love, and established a career, all of which would take place with the memories of previously similar activities. If this were the case, then it might be presumed that the concern to survive has been satisfied.

Williams's account of a series of lives is, as I understand it, a thought experiment to demonstrate what is desired by increased longevity, and what is wrong with increased longevity. The aim of increased longevity is survival of oneself into the future, which requires the belief that one's future self will continue one's present categorical desires. Most people would not believe that they lived longer if increased longevity constituted a series of psychologically disconnected lives. Williams accepts that people's characters change, but he argues that a person will recognise, and act in accordance with the recognition that the changed future character is their character. It will be their character because of the continuity and connectedness of their desires, beliefs and memories, that is, their psychology (p. 93, and Williams, 1976b, pp. 8–9).

What Williams rejects is the possibility that one person could have a series of completely different characters and still be identified as the same person. To emphasise the problem of radical changes in character, he considers the example of Teiresias who changed sex (1973b, p. 94). Teiresias' alterations were complete physical and character alterations, but where he retained his memories. Teiresias, Williams argues, was not a person but a phenomenon. He was a phenomenon because he is based on a fantasy, which must ignore the connection 'between having one range of experiences rather than another, wishing to engage in one sort of thing rather than another, and having a character' (p. 94). In some respects Teiresias resembles the case of intelligent adult reduced to the mental condition of an infant. While the intelligent adult and contented infant may share some memories, they are psychologically different in significant respects. It is for this reason that the intelligent adult would not recognise themself as having survived the accident. In contrast, a series of lives where each life contained the psychology of the previous life would be analogous to the intelligent adult after their accident possessing both their psychology as an intelligent adult and the psychology of a contented infant.

Increased longevity, therefore, requires the continuation of the same character. Undoubtedly a person's character will change to some degree over time. Williams's point is that if someone remains recognisably the same character it will be because of certain fundamental dispositions, which will be reflected in their categorical desires. The tedium of immortality will arise when the categorical desires would 'go away' from life, that a person 'would eventually have had altogether too much of himself' (p. 100). Williams's contention that an extremely long life would become tedious is based on the view that without changes in character people would repeat the experiences of their lives, particularly repeated patterns of personal relationships (p. 90). At first glance, as Nagel (1986, p. 224, n. 3) observes, Williams's hypothesis suggests he is easily bored. Even with a relatively fixed character, the pursuit of a seemingly wide range of activities is possible in a significantly long life span, without getting bored. If life spans were to increase considerably, social and cultural institutions would need to develop to accommodate this change. Although this may require a fundamental modification to social attitudes, if everyone in society experiences increases in longevity, there is no reason why this, and the necessary cultural and economic alterations, cannot occur.

As I understand Williams's argument, his claim that significantly increased life spans would become tedious not only suggests that people will become bored with endlessly pursuing activities and patterns of personal relationships, but more importantly, that this endless pursuit will harden them against the contingencies of life. The activities and attachments that people pursue and maintain will not become more stable and certain, and consequently less fragile. Rather, people will become inured to the consequences and influences that the vicissitudes of their activities and attachments has on their lives. Frequent changes to our activities and attachments will undermine what it is to be committed to a particular person or activity, and consequently lessen the impact of their fragility on us. For example, in Čapek's (1922, p. 240) play, Makropulos cannot remember precisely how many children she has but believes it to be twenty or more. Čapek's point is that once events in a person's life that are relatively rare and special become commonplace, they become reduced in value. Makropulos's emotional coldness arises because she is no longer affected by the vulnerability of the fundamental features of life.

If people are to retain the same character, living for too long will inure them to the fragility of the values that contribute significantly to the good life. As a result, a person's categorical desires will leave them because they will no longer be able to offer a committed answer to questions about the meaning of their existence. The tedium of immortality is not only an issue about boredom, but also about the good life. If, as Williams suggests, a person who has lived too long becomes emotionally cold, their well being will suffer because they will no longer respond appropriately to the vulnerable aspects of their life.

An important caveat must be added to Williams's argument. The tedium of immortality concerns the nature of the good life, and so everyone will experience the boredom of living if they were to live sufficiently long lives. If individuals are responsible for their own lives, however, the length of a person's life before it becomes tedious will be particular to individuals. It may be the case that Jonathan Glover (1977, p. 57), who says that he would welcome the opportunity to live a few million years, would not find living this long tedious. Nevertheless, Glover tacitly accepts that he would want to die: he does not claim to want immortality, that is, a life without end. Williams must accept that different people will desire different lengths of life, and there is nothing in his hypothesis to contradict this. Indeed, if people are to become inured to the vulnerability of life, this must occur over a long period of time. Were people to become so quickly hardened to the fragility of the good life, life would not be very responsive to the contingencies that shape our values.

What my argument has sought to show is the value of longevity. If life is good, more life is better than less, and it is for this reason that death is bad. Death prevents the deceased from experiencing the fulfilment of their categorical desires. These desires provide the grounds for the activities and attachments that we maintain and pursue and which provide our lives with its meaning and value. The fact that one's life is fulfilling makes it worth continuing. Hence, the quality of life is intrinsically bound to the length of life. To increase life spans, therefore, involves promoting a value that is central to the pursuit of the good life.

The tedium of immortality reveals, however, precisely what is objectionable to a technological understanding of Being: that it is possible to have too much control over the vulnerable aspects of our lives. It also demonstrates that the pursuit of longer life spans is not, nor should it be, an attempt to abolish death. Not only would such an endeavour fail but it is also undesirable. To achieve the complete prevention of ageing, were such a goal possible, would not be sufficient to prevent us from dying. Without complete control over every aspect of our lives and the environment, extrinsic factors, such as new variants of diseases, accidents, misadventure and environmental catastrophes would eventually cause each person's death. What the tedium of immortality emphasises, and which is central to the critique of the technological understanding of Being, is that if it were possible to control all aspects of our lives, including these extrinsic factors and the risks they pose, life would eventually cease to be worth living.

3
Justifying the Means

The discussion of the previous chapter shows that, other things being equal, longevity is an important value, and one that is fundamental to the pursuit of the good life. The second area of consideration for assessing the ethical implications of increasing life spans concerns the means for promoting our values. The example of Ritalin, used by students as a means of enhancing their performance in exams, emphasises the need to consider the relationship between our ends and the means for achieving them. In some cases, the means are more important than the ends: the act of climbing a mountain can be more valuable to the climber than reaching the summit (Cole-Turner, 1998, p. 155). When used as a performance-enhancing drug, Ritalin is objectionable because part of the value of good exam results is the effort required in studying to achieve them. What the example of Ritalin demonstrates is that although the end may be good, it does not always justify the means for achieving it. This is particularly so when the way in which a value is promoted involves wrongdoing.[1]

Of concern here will be whether increasing life spans involves wrongdoing, and my focus will be on the use of nonhuman animals and human embryos in the development of biomedicines. As was noted previously, increases in life spans will not be a result of a specific medical treatment, but a range of new medicines that maintain and restore the quality of people's health. Given present medical practices, nonhuman animals (henceforth, animals) will continue to be used as standard research tools for the development of new medicines. In contrast to animal experimentation, human embryonic stem cells offer considerable potential as a future medicine, which could prolong life. In both cases, of present medical practices and of a future technology, living beings will be made to suffer and be sacrificed for the benefit of others.

50

Whether or not this is justifiable will depend upon our beliefs about how we ought to behave towards animals and human embryos.[2]

Animal experimentation

Animals have been used in medicine for over 2,000 years, and they are now a standard tool for research and the development of new medical treatments. What makes animal experimentation controversial is that it usually involves making animals suffer and destroying them. There have been a number of attempts to argue that animal experimentation is morally impermissible. I will outline the ideas of Peter Singer and Tom Regan, where my focus will be on their claims that animal suffering is of equal importance to that of human suffering. My criticism of their arguments will provide the basis for my own objection to animal experimentation.

In order to simplify my discussion, I make two assumptions. First, I assume that vertebrates are capable of suffering. It cannot be proven beyond doubt that they can suffer, but their similar nervous systems, demonstration of pain behaviour and evolutionary theory give good grounds for believing this to be the case.[3] This is not to claim that all non-vertebrates are incapable of suffering. There may be some species that can but the evidence for this is less clear. My argument, therefore, only applies to vertebrates, but these constitute a significant proportion of the animals used in medical research. Second, there have been many examples of medical treatments developed and tested on animals that have proved to be detrimental to human health, such as the use of thalidomide (Singer, 1995, p. 57). The difficulty of applying data from animal experiments to human beings raises questions about its efficacy, which may prove to be sufficient reason to doubt its permissibility. I assume that, when properly interpreted, animal experimentation can be beneficial for the development of medicine. My aim is to question the moral permissibility of animal experimentation even when it is beneficial.

Singer's objection

Singer (1995, p. xii) regards his argument against animal experimentation as an extension of the liberation movements against racism and sexism, which discriminate on the basis of arbitrary characteristics. The condemnations of racism and sexism originate in the belief that human beings are fundamentally equal and should be treated as such. Human beings are not all alike; we differ in our stature, intelligence and personality, for example. Consequently, Singer argues the principle of equality 'is a prescription of how we should treat human beings' (p. 5, originally emphasised). Nonetheless, there must be some basis, one or more characteristics about human beings that provide grounds for the prescription for equal treatment. These grounds are the fundamental interests that all human beings share, such as the need for food, shelter and the fulfilment of one's desires.

A necessary condition for the possession of interests is the capacity for suffering and enjoyment. To illustrate this point, Singer (1993b, p. 57) observes the crucial moral difference between a rock and a mouse. There is nothing that can be done to a rock that will affect its welfare because the rock lacks the capacity for suffering, and so it lacks interests. In contrast, kicking a mouse will cause it pain, which does harm its welfare, and so kicking a mouse is not in its interests. On this basis, many animals, most notably vertebrates, have fundamental interests because they have the capacity for suffering. As a result, the equal consideration of interests must be extended beyond the sphere of human beings to encompass them.

Singer does not claim that the suffering of human beings is always comparable to the suffering of animals. There are certain differences between, for example, an adult human being and an adult mouse, which mean a human being could suffer more than a mouse. A normal adult human being is more self-aware, has greater powers of anticipation, a more detailed memory, has meaningful relationships, can plan for the future, and has a greater knowledge of what may be happening than an adult mouse. In some situations, however, a lower self-awareness may lead some animals to suffer more than fully self-aware human beings. The complexity of the nature of suffering means the comparison between different species will be difficult to make (1993b, p. 60; 1995, p. 19).

Singer also distinguishes between the moral value of beings, where the ability to suffer affords some value to a being. Of more importance is the ability to reason and be self-conscious, which are necessary for possessing future-orientated desires. On this basis, a fully self-conscious and rational adult human being has greater moral value than an adult dog because of the greater complexity of the human's desires. The difference between the moral value of beings might be important where a choice must be made between painlessly killing two beings of different values, but in the case of animal experimentation, what matters is the suffering that the experiments cause those animals that are involved (Singer, 1993b, ch. 4).

The prescription of equal treatment is based on the capacity for suffering of beings. If a choice has to be made between preventing the suffering of an adult dog and that of an adult human being who is fully self-conscious, in most instances the adult human being would be chosen. The choice is based on the greater capacity for suffering of the human being and not a prejudice in favour of human beings. To choose in favour of the human being simply because they are human would represent a 'speciesist' view, where '[t]he term "speciesism" refers to the view that species membership is, *in itself* a reason for giving more weight to the interests of one being than those of another' (Singer, 1980, p. 59, emphasis original).

Speciesism is a form of prejudice akin to sexism and racism because the decision to use an animal rather than a human being in medical research is based on the arbitrary feature of species membership, and not the morally relevant feature of the capacity to suffer. Singer (1993b, pp. 65–8) points out that the principle of equal treatment might lead to the conclusion that it is better to choose a human being than an animal for certain medical research. An ape, such as a chimpanzee, may possess a greater capacity to suffer than a severely mentally disabled person because it is more self-conscious than the person. Should a hypothetical situation arise whereby a single medical experiment on an animal or a person could save thousands of people, to use a chimpanzee rather than a severely mentally disabled person because the chimpanzee is an animal would be speciesist. Singer does not advocate using severely mentally disabled people in medical experiments. What he argues is that we should only consider using animals for medical purposes if we would also be prepared to use human beings with comparable interests.

A rights-based objection

Singer's argument does not provide an absolute objection to animal experimentation. His support for a utilitarian account of normative ethics means he would accept some experiments on animals, should the benefits be sufficiently great to justify the suffering they involve. The purpose of questioning whether we would accept using a mentally disabled person in place of a chimpanzee is to query whether the present benefits of animal experimentation are sufficient to justify them, and he argues that they are not.[4]

While Regan is generally supportive of Singer's argument for extending the equal consideration of interests to include many animals, he is critical of utilitarianism. In particular, Regan (1985, p. 82) is critical of the way utilitarianism fails to value individuals but instead values the satisfaction of interests that people have. The fact that, collectively, one's activities and attachments make one's life what it is and provide it with its distinctive pattern is of no concern to the utilitarian. What is of concern when judging whether an act is right is its consequences for all those interests that it impacts upon, no matter whose they are.[5] The best way, Regan argues, to protect individuals is to recognise them as possessing rights, which are 'trumps' protecting the rights bearer against being sacrificed as a result of a utility calculation (Regan in Cohen and Regan, 2001, pp. 153–4). Regan proposes an account of animal rights that entails the abolition of animal experimentation (p. 127).

While advocating animal rights, Regan is critical of contractualist theories of rights. He argues that, according to contractualism, rights are established by a contract that 'individuals voluntarily agree to abide by' (Regan, 1985, p. 80). Only moral agents, who are rational, autonomous and self-legislating, are directly covered by the contract of rights because only they are able to conceive, understand and adhere to it. The contract cannot directly encompass animals because they are incapable of understanding and respecting it. Nevertheless, according to contractualism, moral agents may have indirect obligations towards animals based on their direct duties. For example, moral agents have a duty not to harm animals that belong to or are cared for by other moral agents because to do so would infringe these agents' rights.

Regan (1985, p. 81) points out that there are many human beings, such as infants and the severely mentally disabled, who are incapable of respecting other people's rights. The contractualist account of rights implies that moral agents only have indirect duties to such people. Yet, Regan argues, to torture an infant, for example, would not be wrong only if it were to upset a moral agent but because it wrongs the infant. Moral patients are those people, such as infants and the severely mentally disabled, who have rights directly but are not capable of respecting other people's rights. They, along with moral agents, have direct rights because they are inherently valuable. Those beings who possess inherent value do so equally, which prevents prejudice and discrimination because no person is less valuable than any other (p. 84).

What provides people with inherent value is the fact that we possess an experiential nature. Human beings are distinguishable from plants because we have 'feelings, beliefs and desires' (Regan in Cohen and Regan, 2001, p. 201). As the distinction between moral agents and moral patients emphasises, the experiential capacity of human beings can vary considerably. Regan therefore uses the phrase 'subject-of-a-life' to refer to all beings that are capable of some degree of conscious experiences. Many animals are also capable of conscious experiences and as such are the subjects-of-a-life, which means they are also inherently valuable and therefore possess rights. He does not claim that animals have the same rights as moral agents but that they do have basic rights, such as the right to life, the right to liberty and the right to bodily integrity. To argue that human moral patients have these rights but that animals which are also the subjects-of-a-life do not, is he claims, speciesist (pp. 170, 290–7).

A problem with the concept of 'speciesism'

One criticism of Regan's account of animal rights is that it is questionable whether only beings that are the subjects-of-a-life possess inherent value. Ecosystems or pieces of art may also be inherently valuable, but they do not possess rights (Warren, 1987, pp. 91–2). A broader criticism of both Singer's and Regan's theories concerns their claims that to fail to extend the equal consideration of interests to animals is speciesist. Williams observes that, '[t]he word "speciesism" has been used for an attitude some regard as our ultimate prejudice, that in favour of humanity. It is more revealingly called "humanism", and it is not a prejudice. To see the world from a human point of view is not an absurd thing for human beings to do' (1993, p. 118).[6]

Williams is critical of theories that seek to occupy the position of an Ideal Observer, which provides a morally objective perspective of the world that is free of all prejudice (pp. 75–7). For example, the fundamental feature of utilitarianism is that it judges moral actions according to the effects they have on maximising the general welfare. In the assessment of the aggregate of welfare, the assessor must consider equally all possible welfare concerns, and this requires that the assessor occupy a position that is concomitant with the Ideal Observer and therefore free of all prejudices in favour of or against any one, or a group of beings capable of having interests. The utilitarian must consider the interests of human beings and animals to be equal because the fulfilment of these interests contributes to maximising the general good.[7]

The position of an Ideal Observer, however, is not one that any person can occupy because it would require stepping outside of the conceptual scheme that grounds our beliefs and values. Part of this conceptual scheme is morality, which is grounded upon fundamental facts about human beings, and it is only by understanding human relationships in respect of these facts that we can come to understand the demands of morality. As Rosalind Hursthouse (1987, pp. 238–9) argues, morality is 'about' human beings because it is grounded upon specific human characteristics. This does not mean that human beings do not have moral relationships that extend beyond the human sphere, but, as Williams contends, '[a] concern for nonhuman animals is indeed a proper part of human life, but we can acquire it, cultivate it, and teach it only in terms of our understanding of ourselves' (1993, p. 118). It is only by understanding moral relationships between each other that we can grasp what is morally significant about our relationships with animals.

The fact that we have moral relationships with animals does not mean they possess rights. The justification of speciesism maintains that morality is 'about' human beings because it is grounded upon facts about people. Rights, along with duties, obligations and claims, represent moral relationships and identify what constitutes morally appropriate behaviour between people. To reflect upon the moral relationships between moral agents and human moral patients (as Regan and Singer do) will suggest the appropriate moral concerns that people should have towards animals. It does not follow from reflecting on these relationships, however, that if human moral patients have rights some animals must also possess rights. Rights represent human relationships and identify appropriate moral behaviour between people; they do not indicate human moral concerns to other beings in the world.[8]

To be speciesist in favour of human beings is not an unjustifiable prejudice but a perception of the world that human beings cannot avoid. Moreover, this preference towards human beings may also justify deciding in favour of a human being to the detriment of an animal in a situation where Singer would claim that the animal has a greater capacity to suffer or was of greater moral value. When a choice must be made, our understanding of moral relationships will determine what course of action in a particular situation is the right one. Nevertheless, the acceptance of humanism, as Williams describes it, does not necessarily justify animal experimentation.

An integrity-based objection

An alternative to Singer's and Regan's objections to animal experimentation builds on Williams's observation that we can cultivate our response to animals by reflecting upon our relationships with other people. Singer and Regan point to an inconsistency in our attitude to human beings and animals with comparable capacities for suffering. While society is prepared to use animals in medical research, it is not prepared to use human beings for similar purposes. Singer and Regan advocate the equal consideration of interests and claim that animal experimentation represents a speciesist attitude. To be speciesist is not a prejudice but animal experimentation does point to an inconsistency in our moral outlook that is difficult to justify.

In his argument against utilitarianism, Williams (1973a, pp. 98–9) provides a thought experiment concerning a botanist called Jim. Jim finds himself in a village in South America where Pedro, the captain of a group of armed guards, has ordered that 20 innocent Indians be killed. Pedro considers Jim to be an honoured guest and proposes that if Jim executes one of the Indians the remaining 19 will go free; if Jim does not do this, all 20 will be shot. In accordance with utilitarianism, the right thing for Jim to do is kill one of the Indians because saving the lives of 19 people outweighs the detrimental consequences of killing one person. Williams argues that, for Jim, killing anyone will undermine the integrity of his own projects, which we are left to presume do not involve killing others.

By raising the issue of Jim's integrity, Williams criticises the utilitarian failure to distinguish between individuals and their commitment to and identification with the projects (their activities and attachments) that shape and define their individual lives (pp. 116–17). The issue that Williams raises is the extent to which one person is required to sacrifice their commitment to their activities and attachments for the sake of other people and to produce some good that is in part determined by others. He concludes that, in this case, Jim should 'probably' sacrifice his projects, but maintains that utilitarianism does not describe the complexity of this decision (or the regret that Jim might feel) (pp. 102–6). Whether or not we agree with Williams's conclusion that Jim should kill one of the Indians, we can agree with his argument that utilitarianism cannot adequately explain why, if at all, a person might be required to sacrifice their projects, activities and attachments.[9]

While I agree with Williams's criticism of utilitarianism, his reference to a person's integrity is incomplete. Williams criticises utilitarian calculations for alienating a person 'in a real sense from his actions and the source of his actions in his own convictions' (p. 116).[10] Nonetheless a problem for Williams's critique is not his account of a person's integrity but the nature of the convictions on which their integrity is based. For example, a utilitarian demonstrates their commitment to utilitarianism when they willingly abandon their personal projects because of the requirements of a utility calculation. Moreover, there are many examples in history of people demonstrating considerable integrity in their convictions when pursing immoral ends, such as the devoted followers of Adolf Hitler or Josef Stalin who zealously pursued their leaders' policies. A situation might arise, therefore, where alienating a person from their convictions is justifiable because their acting upon them would involve wrongdoing.

A way of understanding Williams's idea of integrity is to conceive of it in relation it to the pursuit of the good life. In accordance with such an understanding, the pursuit of one's activities and attachments must contribute to (or at least not detract from) living well, not just for the individual whose activities and attachments these are but also for the members of the society in which this person lives. Indeed, the nature of our relationship with others and with the world in which we live constitutes an important aspect of the good life and what makes life go best. Central to our relationships is morality, and to pursue the good life requires that we act in a morally appropriate way. Our convictions about what constitutes morally appropriate behaviour circumscribe and identify what people should do and what we can expect and demand of each other. The integrity of our convictions demonstrates our commitment to behaving appropriately and in a way that contributes to living well. Actions that undermine the integrity of a person's moral convictions alienate them from their pursuit of the good life. It is by understanding how these convictions contribute to the good life in general that we can comprehend when, if at all, it is justifiable to interfere with a person's integrity. What is objectionable about animal experimentation makes clearer this relationship between the integrity of our moral convictions and the pursuit of the good life.

Animal experimentation undermines the integrity of our convictions about the infliction of suffering. Singer and Regan rightly observe that animals, specifically vertebrates, have in common with human beings certain fundamental interests, not least the avoidance of suffering. The comparisons they draw between human moral patients and many of the animals that are used in animal experimentation usefully illustrate this point. The fact that we would not tolerate the use of infants or the severely mentally disabled in medical experiments is not because they are human beings. The subject of their species membership does not arise, except in philosophical discourses on the merits of animal experimentation. What prevents the use of human moral patients is the suffering they would experience, which in part, grounds our obligations to them.

There are situations where causing a person to suffer is morally permissible. For example, a surgeon may make their patient suffer as a consequence of an operation. So long as the operation is intended to benefit the patient and their suffering is not the result of malicious or negligent behaviour, the surgeon's actions are morally permissible. There are also situations where people experience suffering during medical experiments that serve to benefit society as a whole and not simply the experimental subjects. It might be thought that what is central to such cases is that the patients and experimental subjects consent to the procedures that make them suffer. This is not always the case, with people who are generally capable of giving consent undergoing emergency surgery when they are temporarily incapable of so doing, or parents who consent to operations for their children. What is common to these situations is that any suffering results from actions that aim at meeting an individual's fundamental needs and interests or at least do not threaten them.[11]

This is not intended to be an exhaustive list of the different contexts in which making another person suffer can be morally permissible, but to demonstrate the type of characteristics that are salient in such situations. It is by reflecting upon the situations in which we would permit making a person suffer, and why this is so, that we can come to know when, if at all, making animals suffer is permissible. The fact that society does not nor would it permit the use of human beings for many of the experiments in which animals are used raises the concern that we are behaving or tolerating activities that are inconsistent with our convictions about what constitutes appropriate moral behaviour. Situations do arise that compromise our convictions, such as the hypothetical case of Jim and the Indians. Although Jim believes it to be wrong to kill, he might consider it better in this particular situation to kill one Indian to save 19, but in so doing recognise and regret the compromise he is making. The fact that Jim concedes that in this specific case he ought to do what he otherwise believes to be wrong challenges the integrity of his convictions, but it need not undermine them. The understanding of how we should respond to moral conflicts without undermining the integrity of our beliefs is fundamental to moral behaviour.

In contrast to Jim's moral conflict, animal experimentation is not a response to a specific situation, such as Singer's hypothetical example of an experiment on a few animals that could save thousands of people. It might be that were such a situation to arise, experimenting on a small number of animals would be morally permissible and so might not threaten the integrity of our convictions, although, as with Jim's actions, this is questionable. Animal experimentation, however, does not involve the use of a small number of animals to find a specific cure of considerable benefit. Millions of animals are used each year as standard tools of medical research and development.[12] It is the systematic mistreatment of animals in order to develop medicines that undermines the integrity of our convictions about when causing suffering is morally permissible.

To maintain and restore good health, and as a consequence increase longevity, is a valuable endeavour, one which will require greater control over our bodies and environment. A technological understanding of Being is one which seeks too much control and in so doing threatens the pursuit of the good life, and it can explain why animals are used as a resource, as 'interchangeable and anonymous objects (Arluke, 1992, p. 33) in the development of medicine. The way in which animals are used is not demonstrative of a harmonious relationship. A harmonious relationship is, I have proposed, one that promotes the good life, which animal experimentation, in undermining the integrity of our moral convictions, does not accomplish.

What implications abolishing animal experimentation would have for the quality of our health and longer life spans is unclear. There are alternatives to using animals in medicine, and although I have assumed that there are benefits to animal experimentation, these may be overstated. What is clear from my argument is that viewing and using animals technologically challenges the pursuit of the good life.

Embryo experimentation

Any increase in average life spans beyond present expectations will be the result of biomedicines that can repair the damage that causes the ageing process. In recent years, stem cells have become the focus of much interest because of the potential they offer for medicine. Stem cells might be used to replace damaged cells and to develop new tissues, while human embryonic stem cells offer the potential for developing immune-compatible cells and tissues. What is more, the self-renewing nature of some stem cells means they can provide a ready supply of replacement cells and tissues.

While they may prove to be medically important, stem cells are also controversial for a number of reasons, but particularly because their production can involve the destruction of human embryos. Unlike other sources of stem cells, human embryonic stem cells offer the greatest potential for medicine because they are pluripotent and can become any tissue in the human body. The use of cell nuclear replacement techniques to produce embryos from soma cells also means stem cells could be produced that are immune-compatible. It is the issue of the moral permissibility of destroying human embryos that will be the principal focus of my discussion. My argument will, however, explain why creating embryos specifically so that they can be used and destroyed for medical purposes is morally permissible.

The debate on using embryonic stem cells focuses on the moral stature of the embryo in the first two weeks after fertilisation or creation by cell nuclear replacement. I refer to the embryo during this stage of development as the early embryo. It is only during the blastocyst stage of embryonic development, approximately six days after fertilisation, that it is possible to obtain stem cells from the early embryo. The limit of 14 days is morally significant because it is at approximately this time that the primitive streak, a line of cells forming the basis for the embryo's future neural network, develops. Until the development of the primitive streak, the early embryo can divide into two or more embryos, and if the embryo has not implanted into the uterine wall by this stage, it will not survive. The moral significance of these features of the early embryo will become apparent as the discussion progresses.[13]

The moral stature of the early embryo

Whether destroying early embryos in order to obtain their stem cells is morally permissible is a question of their moral status. For an entity to possess moral status is for it to be an object of moral concern such that there are ways in which it should or should not be treated. There are three reasons for denying that the embryo lacks moral stature: it is not a person, in the philosophical sense; it is not a human being; or because it lacks interests. I will assume, for the moment, that the possession of moral stature is sufficient to prevent an entity from being destroyed.

Personhood and moral status

An initial reason for denying that the early embryo lacks moral status is because it is not a person, in a philosophical sense. In order to avoid confusion, I will use Person to identify the philosophical sense of the term 'person', with the lower case denoting the synonym for human being. A Person is typically a rational, self-conscious being, although other psychological characteristics are sometimes included as criteria for the definition.[14] Of central importance to the debate about what entities possess moral stature is explaining why certain characteristics are morally salient (Marquis, 1989, pp. 185–8). The psychological characteristics associated with Personhood are necessary for a being to be able to discern what is right or wrong about its treatment and the way it should behave towards others. A further advantage of the Personhood account of moral stature is that it is not restricted to human beings. Any being that is rational and self-conscious will have moral stature, and it has been argued that some animals, such as the great apes, possess these characteristics.

While Personhood may not be restricted to the human species, it also does not apply to all human beings. Newborn infants, for example, are not Persons and so lack moral stature on this account, which is not normally the case. What is more, the addition of more criteria for Personhood can further restrict its scope to exclude more people and animals that we generally believe possess moral stature (Steinbock, 2006, p. 28). If the concept of a Person does not encompass all those entities that have moral status, it remains possible that the early human embryo is an object of moral concern.

The early embryo is not a human being

While the early human embryo may not be a Person it would seem obviously to be a human being. As a small collection of cells, the early embryo is so unlike anything we think of as constituting a human being that this can be denied (Glover, 1977, p. 124). Williams (1988b, p. 221) compares the differences between an early embryo and an adult human being and the differences between an acorn and an oak tree. He claims that just as people would not naturally refer to an acorn as an oak tree, so most people do not naturally consider an early embryo to be a human being. The term 'pre-embryo' is sometimes used to label the early embryo to emphasise that it has not reached a sufficient developmental stage to be counted as a member of any species. It may be human, but it is not a human being.

In response to Williams, we can point to H. G. Woodger's observation of a shortcoming in the terminology used in the biological sciences, which tend to 'identify *organism* with *adult*' (1945, p. 95, emphasis original).[15] Williams seems to be guilty of a similar shortcoming. An acorn will belong to a specific variety of the *Quercus* family, such as the Common Oak, as will the fully developed tree. Similarly, an adult person is a fully developed human being and an embryo is a very young human being. The fact that the early embryo consists of a collection of cells, in particular, that these cells are undifferentiated and can become any tissue of the body, does not mean it is not a human being. Rather, these facts suggest our understanding of what a human being is needs to reflect the different stages of human development.

The claim that the embryo lacks moral status because it is not a human being entails the assumption that to be a human being is a necessary criterion for possessing moral stature. The argument concerning animal experimentation indicates that this is not the case, and consequently even if the early human embryo were not a human being it might still be an object of moral concern. To be clear about the argument justifying speciesism, it concerns the loyalty and identification that one has with one's species (Williams, 2006, p. 150). It is not an argument which claims that biological membership of the human species is itself morally significant. Morality is grounded upon certain features about human beings, such as the fact that we are, other things being equal, capable of suffering and have complex relationships, which may involve loyalty to others. While many of these features will be biological, and in some cases will be unique to our species, they are not dependent upon species membership (rather, species membership signifies the possession of certain characteristics). What is more, the fact that other species possess many of these features is, as I have argued, morally significant.

Interests and moral status

A general characteristic of human beings, and other vertebrates, which is morally significant is the fact that we have interests. It is in our interests to avoid suffering and to experience enjoyment, which means those beings with interests are the objects of moral concern. There is a close relationship between our interests and our desires, although not necessarily a direct one. We often desire things that are not in our interests, or do not desire things that are in our interests (Steinbock, 2006, p. 28). Nevertheless, we do desire what is in our fundamental interests, that is, that our life goes well and flourishes. This link between interests and desires implies that only conscious beings can have interests because consciousness is necessary for desires, which in turn implies that only conscious beings can have moral status, thereby precluding the early embryo.

Unlike the idea that being a member of the human species is necessary for possessing moral status, the interests account identifies a feature of human beings, and other animals, which is morally significant, and it can explain why this is so. In contrast to the Personhood account of moral status, the interests account, with some modification, can point to why most human beings have moral status and why early human embryos do not. Without clarification, the interests account still risks denying moral stature to people and animals when they are asleep or temporarily comatose because, in such situations, they are unconscious. The problem cannot be solved by maintaining that the temporarily comatose and sleeping have future interests because this will also apply to the early embryo (Marquis, 1989, p. 188). We can, however, make sense of having dispositional desires, desires that we have although we may not be conscious of them at all times, such as the desire not to be killed (Steinbock, 2006, p. 31). It is not obvious, however, that possessing interests is a necessary criterion for having moral status. As was noted earlier, Regan restricts inherent value only to those beings that are the subjects-of-a-life, but we might consider other, non-conscious entities as having inherent value and, in virtue of this, as possessing moral status. If this is the case, then early human embryos, although lacking interests could still be morally valuable.

The potentiality argument

None of the three accounts for establishing the moral status of a being can show definitively that the early embryo lacks moral stature. What they do reveal is that the embryo lacks certain morally salient characteristics. Nevertheless, as a human embryo, it will, other things being equal, attain them and specifically those features necessary for Personhood. This feature of the early embryo, its potential to become a Person, may be what provides it with moral stature.

The meaning of potentiality

An initial problem with the potentiality argument is that it appears to ground the moral status of the embryo on the future Person it will become and not on any of its actual properties. It is by no means clear how the future Person the early embryo will become can impart moral status upon it because, by definition, a potential Person is not a Person. For example, citizens of the USA are potential Presidents, but they are not Presidents and crucially do not have the same powers as the President because of this fact (Marquis, 1997, p. 99). Moreover, the argument implies that the wrongness of killing early embryos does not derive from any harm caused to the early embryo itself but to the future Person it would have become. To kill an early embryo prevents a Person who would otherwise have existed from coming into existence (Reichlin, 1997, p. 3). Yet, the same argument is also applicable to contraception. It might be responded that the wrongness of killing early embryos, in contrast to contraception, is the fact that the embryo is already in existence. This, however, is to shift the focus of the argument from potentiality, understood as relating to the properties of the future Person the early embryo will become, to the actual properties of the early embryo (Glover, 1977, p. 122).

An alternative understanding of potentiality is to view the early embryo as a possible Person, and it is in virtue of this feature that it has moral status (Reichlin, 1997, pp. 2–9; Perrett, 2000, p. 188). Such an understanding of potentiality would mean that sperm and ova are potential Persons and therefore possess moral status because, in the right circumstances, they will develop into a zygote and eventually a Person. Moreover, given recent developments in modern technology, namely cell nuclear transfer techniques, soma cells are possible Persons because a soma cell could be manipulated into becoming an embryo and consequently a Person. When understood as possibility, the potentiality argument confers moral stature on any entity that can logically become a Person and not simply on the early embryo.

A third understanding of potentiality maintains that the embryo is a probable Person (Reichlin, 1997, pp. 9–12; Perrett, 2000, p. 189). While it is possible that many different forms of biological tissue could become Persons, it is improbable that they will do so. In contrast, it is probable that the early embryo will become a Person. As a consequence, the potentiality argument maintains that the early embryo has moral stature because of the high degree of probability that it will develop into a Person. Knowledge of foetal development means this is not as strong a claim as it might at first appear. The early embryo has only a 25–30 per cent chance of surviving until birth (Singer, 1993b, p. 161). If a high degree of probability of becoming a Person is necessary for possessing moral stature, then the early embryo will not have this status.

Active and passive potentiality

The account of potentiality that best explains what is meant by it is that of Aristotle (2004, IX). He distinguishes between the active potential and the passive potential of a being to transform. The active potential for transformation is intrinsic to a being; it is part of its very nature. In contrast, the passive potential of a being for transformation is dependent upon external forces. For example, an acorn has the active potential to become an oak tree, while the oak tree has the passive potential to become sawdust (Perrett, 2000, p. 189). Aristotle's distinction between active and passive potentiality is a metaphysical claim about the teleological nature of a being and its possibilities. The nature of the acorn is to become an oak tree; the transformation from acorn to oak tree is essential to its existence, and if it does not actualise its potential to become an oak tree, it will die and cease to be. The passive potential of the oak tree to become sawdust is not part of its nature; it will not cease to exist if it does not actualise this potential. The actualisation of the passive potential of the oak tree to become sawdust must be brought about by external forces. In contrast, the actualisation by the acorn of its active potential to become an oak tree is intrinsically driven, with it needing only an environment conducive to this transformation.

Aristotle's account of and distinction between active and passive potentiality resolves the problems of potentiality noted earlier. Active potentiality grounds the moral status of the early embryo in a feature that is intrinsic to it and not the characteristics of the Person it will, other things being equal, become. Moreover, the distinction between active potential and passive potential explains why the possibility of becoming a Person is insufficient to grant soma cells and germ cells moral stature. The possibility of becoming a Person is not an intrinsic feature of soma cells. Yet, it might be argued that the purpose, the telos of a germ cell is to become an embryo and hence a Person. What makes this potential of gametes passive rather than active, as Aristotle (2004, IX, 1049a) notes, is the need for the gametes to 'change'. In order to become a zygote gametes must experience an external influence, a trigger mechanism that unites two gametes together (Oderberg, 2000a, pp. 8–22). The account of active potentiality as an intrinsic feature of the early embryo also resolves the empirical assessment of the moral stature associated with potentiality understood as probability. No matter what probability the early embryo has of becoming a Person the active potential to do so is intrinsic to it.

The probabilities understanding of potentiality emphasises an important feature of the Aristotelian account of active potential. The intrinsic nature of active potential means it must be actualised, in contrast to its passive potential if it is to survive. An oak tree remains an oak tree even if it is not turned into sawdust, but an acorn ceases to be if it does not develop into a tree. An early embryo that cannot actualise its potential for Personhood will, therefore, die, and Massimo Reichlin (1997, p. 6, n. 41) proposes that the failure of such a large proportion of embryos to implant in the uterine wall may be a result of them lacking the active potential for Personhood because of genetic defects. If we accept the teleological implications of the Aristotelian account of potentiality, it will require that the definition of a Person has a wide scope because human beings will develop and survive without being capable of complex reasoning and moral decision-making.

Two problems for the potentiality argument

The Aristotelian account of active and passive potential can provide a coherent account of the potentiality argument, which might explain why the early embryo has moral stature. The argument implies that a human being is, from its inception as a zygote, a unified individual that is ontologically unique, gradually accruing the capacities necessary for Personhood as it develops.[16] Yet, the characteristics of the early embryo contradict this description of it. What makes the stem cells of the early embryo useful to medicine is their undifferentiated nature, which means they can become any cell in the body. The absence of a specified function for the cells, however, makes it impossible to know which cells will form the embryo proper and which will develop into the extra-embryonic material, such as the placenta. What is more, until the development of the primitive streak, the early embryo can divide into two or more embryos. While the potentiality argument might accord the embryo moral status, it cannot do the same for the early embryo. There are a number of counterarguments that can be made against this objection addressing each of these characteristics of the early embryo.

Extra-embryonic material

Until the development of the blastocyst, around six days after fertilisation, the early embryo is a cluster of undifferentiated cells called the morula. These cells will form the embryo proper and the embryonic support systems, such as the placenta and umbilical cord. At the morula stage of development it is impossible to identify those cells that will form the embryo and those that will form the extra-embryonic material. Jason Eberl (2000, p. 144–6), among others, argues that because the early embryo consists of a number of undifferentiated biological entities it cannot constitute a unified individual. Only with the development of the primitive streak, when the embryo has completed implantation into the uterine wall and there is a clear distinction between the embryo's support systems and the embryo proper does the embryo constitute a single biological entity, that is, a unified, ongoing individual.

An initial problem with this argument is that the morula is not a mere collection of unconnected cells, as is implied by some of those who oppose the idea of the early embryo possessing moral stature. Alan Holland (1990, p. 30) points to the fact that the morula is likened to a blackberry, which, rather than implying a loose cluster of cells, suggests a unified biological entity. The objection also rests on the claim that none of the cells of the morula are specialised, thereby making it impossible to know their future function. There is some evidence, however, that communication occurs between the cells, which might indicate that they are in fact specialised to some degree. Even if we grant that the cells are undifferentiated, some co-ordinating principle, presumably coupled with communication between the cells, may still be at work directing the specialisation process, which implies that the morula is a unified individual (Holland, 1990, p. 30; Deckers, 2007, pp. 274–5; Tollefsen, 2001, p. 72).

One or more of the cells of the early embryo must become the embryo proper if it is to survive. As Holland (1990, p. 31) argues, it is not obvious why it is necessary to be able to identify which cells will become the embryo. It is enough to be able to claim that 'I'm-in-there-somewhere'. We can trace our lives back to the initial zygote from which we developed even though we may not be able to identify the precise cells. The objection to the potentiality argument also implies that for a human being to constitute a unified individual there must be a clear boundary between it and its environment. The boundary of an adult human being is vague. We shed skin cells and hair, and our intestines contain bacteria, all of which make it difficult to delineate between where a human being begins and where they end. There is no reason to suppose that this should be any different for very young human beings, where it may be difficult to discern between the early embryo proper and its extra-embryonic material (p. 32).

The problem of twinning

A seemingly more difficult problem for the potentiality argument is that of twinning, because the ability of the early embryo to divide into two or more embryos raises some specific logical puzzles. When the early embryo twins, it divides into two new embryos, each of which is identical to the original embryo. (The embryo can divide into more than two embryos, but I assume that it twins for simplicity's sake.) The process of twinning violates the principle that no two objects can occupy the same space at the same time: because they have the same origins, the twins will have existed at the same time as the original embryo. What is more, the principle of transitivity, that if a=b, b=c, then a=c, means both new embryos are one and the same embryo (Holland, 1990, pp. 33–4).

The potentiality argument implies that the embryo is a unified individual with an ongoing ontological identity, and this is achieved when the embryo is no longer capable of division. A solution to these logical challenges, therefore, is to claim that the embryo becomes a potential Person only with the development of the primitive streak, when it can no longer divide into twins. It is unclear how this resolves the logical problems. An advocate of this solution must explain the relationship between the early embryo and the embryo 'proper', and in a way that does not give rise to the logical implications of twinning.

The difficulties become clear by focusing on the single-celled zygote, Z, and its division into two cells, X and Y, each of which are separate zygotes and continue to develop into Persons. The logical problems arise because X and Y are identical to Z. One possible solution is to deny that X and Y are identical to Z, but rather come into existence at the moment of division, which is also the moment when Z ceases to exist (Persson, 2003, p. 510).

In support of the potentiality argument, and against the idea that Z ceases to exist when it divides, it might be argued that the principle of transitivity does not imply that 'if X *was once* Z and Y *was once* Z, they must *now* be the same' (Deckers, 2007, p. 279, emphasis original). We can make sense, from our own experiences as maturing and developing individuals, of an entity retaining its identity over time while also experiencing some degree of change. Nonetheless, as was observed in the previous chapter, we can accept that entities change qualitatively over time while retaining the continuity of their identity. It will be the case that as X and Y develop they will change qualitatively and no longer resemble Z. This does not solve the problem. On the one hand, if they retain their identity with Z, the problem of transitivity remains unchallenged. On the other hand, if the qualitative change is so radical that X and Y are no longer identical with Z, Z must cease to exist when it divides.

A more obvious problem for the explanation of the relationship between Z and X and Y is the strangeness of the idea that Z ceases to exist upon division. The strangeness occurs because X and Y must originate from something, and the obvious candidate is Z. If Y were to die immediately after Z divides, X would be identical to Z. The same would be true of Y if X ceased to exist the moment after Z divides. In both cases there would be no question of Z ceasing to exist upon giving rise to either X or Y (Priest, 2000, pp. 68–9).[17]

The example of either X or Y dying at the moment of division points to another peculiar aspect of claiming that human beings begin life with the development of the primitive streak when they become a unified individual. Although this may be necessary for twins, it is unnecessary for embryos that do not divide. If we could trace the life of individuals back to their origins, we could trace a non-twin, as a unified individual, back to the moment they were a single-celled zygote. At no point does their existence raise the logical problems of existing in the same place as another individual or violating the principle of transitivity, because they do not twin. In contrast, so as to avoid these logical problems, it would only be possible to trace the lives of twins as unified individuals back to the development of their primitive streaks, creating an asymmetry in the biological development between twins and non-twins (Holland, 1990, p. 33).

The few responses to the difficulties of extra-embryonic material and the logical dilemmas considered here are by no means a conclusive response to opponents of the view that the early embryo lacks the potential to become a Person because it is not a unified individual with an ongoing ontological identity. Nonetheless, the responses are sufficient to raise doubts about the objections to the early embryo possessing active potentiality. In part, the responses point to broader difficulties that extend beyond concerns about the moral status of the early embryo. The logical dilemmas arising from the embryo's ability to divide are not confined to the embryo, with similar problems arising, for example, with amoebas. The responses also point to the need to accept certain features about human beings, namely that the concept of a human being is not synonymous with an adult person. In so doing, it is important to recognise that the claim that human beings are unified individuals is not as determinate as it might imply. The point at which we begin and cease to exist are indeterminate; fertilisation and death are processes (Becker, 1975). In a similar way, the boundaries of a human being, be they an early embryo or an adult, are also indeterminate.

Active potential and moral status

There are two remaining issues that the argument for the potential of the early embryo to become a Person must address. First, the argument must explain why the potential to become a Person is morally significant; and, second, it must clarify how much moral status the early embryo's potential grants it.

The account of active potentiality maintains that, as human beings develop, they gradually actualise their potential to become a Person. This process of actualising provides human beings with those capacities that are necessary for Personhood. Roy Perrett (2000, pp. 192–3) questions the moral relevance of this natural process. The potentiality argument is a metaphysical claim about the purpose of the embryo; it does not identify biological characteristics, such as the capacity for suffering, or psychological features like rationality, that are morally relevant.

Nevertheless, the potential to become a Person identifies a biological process, one of developing various biological features that are morally relevant, where the end of this process is Personhood. The early embryo possesses moral stature because it represents the beginning of a process, with a morally significant end. Yet, the fact that the early embryo is at the beginning of this process and lacks morally salient features also points to an answer to the second question about the degree of moral stature that it possesses. At best, the potential to become a Person grants the early embryo sufficient moral stature to require that it be treated with respect. A sign of respect in this context might involve using an early embryo with active potentiality only for a worthwhile medical purpose and where there are no alternatives that can achieve the same end. A worthwhile medical use of an early embryo would be any activity that leads to an improvement in knowledge or medical treatment that restores or maintains the quality of health.[18]

Creating embryos and moral status

The account of the active potential of the early embryo also explains why there is no moral difference between using spare embryos from infertility treatment and creating embryos specifically for their stem cells. One argument against creating embryos for medical purposes is that it is to use an intrinsically valuable entity merely instrumentally. This is in contrast to spare embryos, which although used instrumentally, are created because of their intrinsic value and with the purpose that, other things being equal, they will develop into adult human beings.[19]

A counterargument maintains that it is better to create embryos specifically for medical purposes rather than use spare embryos because, unlike spare embryos, they will never be in a position to actualise their potential to become a being with interests (Steinbock, 2006, p. 33). This latter argument involves an understanding of potentiality as probability while the former argument determines the moral status of the embryo on the basis of the intentions for its creation and not on any features of the actual embryo. The account of active potentiality that I describe grounds the moral stature of the early embryo on an intrinsic feature of the embryo. As a consequence, spare embryos and those created specifically for medical purposes have the same degree of moral status, and neither type of embryo is preferable to the other when used in medicine.

The connection between potentiality and the process of actualisation implies that an early embryo that cannot develop into a Person because of an inherent flaw does not possess active potentiality and so does not have moral status. This would mean that there are no moral implications to creating early embryos that are incapable of implanting into the uterine wall in order to obtain their stem cells (Alexander Meissner and Rudolf Jaenisch (2006) claim that such embryos are feasible). Indeed, embryos created using cell nuclear replacement techniques are thought to have an extremely low probability of implanting compared with those produced from a sperm and an egg because of inherent flaws arising from the transfer technique (Jaenisch, 2004). Nevertheless, a low probability of implantation does not mean the early embryo lacks the active potential to become a Person. In order to deny that embryos created using cell nuclear replacement techniques possess active potentiality, it must be shown that they could never develop because of an inherent flaw.

The use of early embryos in medicine, and in particular, the creation of embryos specifically for this purpose, would appear to be indicative of a technological understanding of Being. Human beings are being used as resources for gaining control over a vulnerable aspect of our lives, specifically, our health and longevity. In contrast to animal experimentation, so long as early embryos are treated respectfully, and especially if they lack active potentiality, their use in the development of stem cells is morally permissible and does not challenge the integrity of our moral convictions.

The claim that using animals as a resource is representative of a technological understanding of Being but that using early human embryos for similar purposes is not might appear contradictory. It may be comparable to Heidegger contrasting a windmill with a hydro-electric power station on the Rhine. They are both human constructions using natural and renewable sources of energy for human ends. The technological understanding of Being, as I have described it, cannot distinguish between them. Nonetheless, we can make sense of the claim that a technological understanding of Being does not represent an approach to the world, including human beings, that is conducive to the pursuit of the good life. By challenging our moral convictions and commitment to behaviour that contributes to living well, animal experimentation alienates us from the pursuit of the good life. This is not the case with using human embryos in medicine. Their potential to become a Person only provides them with sufficient moral stature to require that they be treated respectfully, which using them to maintain and restore the quality of people's health, when other means cannot achieve the same end, accomplishes.

4
Longevity and the Problem of Overpopulation

The third aspect of increasing life spans to consider is their consequences, and my focus will be on the impact of extending longevity on the availability and distribution of resources. The global population is already expanding at a remarkable rate. Between 1930 and 1986 the world's population expanded from two billion to five billion people and is now nearing seven billion people (Battin, 1998, p. 149; Guillebaud and Hayes, 2008). Population growth has traditionally been the result of high birth rates, with the control of population expansion focusing on reducing these rates. Increases in longevity, however, introduce a new dimension to controlling population levels. Substantial increases in longevity will inevitably mean a considerable reduction in mortality rates, which in turn could provide a new source of population growth.

To simplify my argument, no account will be made of actual population levels, global disparities in population growth or the effects of immigration on population sizes. My concern is only with the relationship between birth and mortality rates and their effect on population growth, and I assume these are the only influences on population size. As throughout my argument, I assume that increases in longevity will apply to all people in society and that resources will be distributed equally, assumptions I later challenge. Overpopulation is a problem concerning the distribution of resources. I will discuss resources in general, but my argument applies to specific resources such as healthcare, which I consider in more detail in the following chapter.

Preliminary aspects of population theory

In what follows I outline certain assumptions and arguments that are necessary for any assessment of the impact of increasing life spans on population levels. They concern the possibility of overpopulation; the relationship between population levels and the distribution of resources; the limitations on preventing overpopulation; and an assessment of the alleged need to justify an asymmetry in our obligations relating to bringing an individual into existence.

The problem of overpopulation

A central problem for population theory is whether there is in fact the possibility of overpopulation. I assume that there is. A finite world can have only finite resources, which would seem to entail that only a finite population can be supported. Overpopulation is undesirable because, as Thomas Malthus (1798) argues, it would lead to famine, the spread of disease and war, the consequences of which would be a cruel reduction of the population. More recent problems are that overpopulation also contributes to global warming as larger populations consume greater quantities of natural resources and produce more greenhouse gas emissions.

In response to Malthus, it has been argued that resources are socially defined, and that technology can be used to support a large population (Notestein, 1970, p. 382). Such claims are less convincing than the Malthusian argument. No matter how they are defined, when considered diachronically finite resources cannot support a potentially unlimited population. Technological developments may help, but, as was observed earlier, Jonas points out that technology often begets as many problems as it is utilised to solve. Moreover, the use of natural resources and the impact of population growth will have an effect on the natural environment that may be undesirable. For instance, although wind turbines are beneficial because they utilise a renewable source of energy, their size and location mean their use might be restricted by aesthetic values, their noise, and concerns for their impact on wildlife. The promotion of aesthetic values, and of valued environments, contributes to the good life. Furthermore, while there may be sufficient land to increase city sizes, their expansion will impact on the natural environment, reducing essential agricultural resources and destroying vital eco-systems.

Resources and the quality of life

In supporting the Malthusian view, a second assumption must be made that there is a strong relationship between resources and the quality of life. With only finite resources, the size of a population will determine how many resources are available for each individual; and the amount of resources available for each individual will be a determining factor in the quality of their lives. Overpopulation occurs when the addition of an extra person to a given population reduces the quality of life for every individual below some level.

When considering the relationship between resources and the quality of life, a distinction must be drawn between real overpopulation and nominal overpopulation. Real overpopulation occurs when the addition of an extra person to the population of a given society removes the opportunity for all members of that society to flourish. The real optimal population level exists when there are sufficient resources to ensure that every individual has the opportunity to flourish. The nominal optimal population level depends upon a given society's values and how these contribute to that society's conception of the good life. Fundamental to the idea of a nominal optimal population is a pluralist view of the pursuit of the good life: there are a number of ways in which people can and do flourish. The nominal optimal population level represents a socially and culturally preferred number of people in a given society, ensuring each individual has sufficient resources to have the opportunity to flourish according to that society's ideal of the good life (an ideal that is provided by the cultural background).[1] Nominal overpopulation occurs when the addition of an extra person to a given population reduces the ability of every member of that society to pursue a socially preferred view of the good life. There will be sufficient resources available to retain the opportunity to flourish, albeit for an ideal of the good life that promotes slightly different values.

A number of interrelated aspects about this distinction should be noted. Much of the literature on population theory considers the quality of life in terms of life being worthwhile.[2] Whether one's life is worthwhile, that is, worth living, is highly subjective. Richard Hare (1988, p. 75), for example, considers his life, for the most part, to have been worthwhile during his internment in a Japanese prisoner of war camp. Only when he became a slave building the Burmese railway does he believe that his life was not worth living. What sustained him during this period of slavery was his belief that his situation would eventually improve. He needed no such belief to consider his life worth living during his internment, although he would have preferred a marked improvement in his conditions. Some people might disagree with Hare about conditions similar to his internment and require the belief that their situation would improve in order to want to continue living. My concern in describing population levels in terms of the opportunity to flourish is to emphasise the objective conditions for assessing whether the addition of an extra person to a given society causes real or nominal overpopulation, where a certain amount and type of resources are required in order to flourish.

While there is a strong relationship between available resources and the quality of life, it is not a direct relationship. There is a limit to the amount of resources that any one individual can utilise. Furthermore, the fact that sufficient resources exist for someone to flourish does not entail that they will flourish; and an individual's life may still be worth living, and be of considerable quality, even if they do not flourish. What my argument implies, however, is that where there are insufficient resources to provide the opportunity for flourishing (in any of its various forms), life would not be worth living. As a consequence, defining the real optimal population level in terms of the opportunity for human flourishing requires better conditions than are necessary for life to be only just worth living.

Some people might claim that Hare's life during his internment resembles what the quality of life would be like at the real optimum population level. This does not concur with my understanding of what gives life its meaning and value. The pursuit of the good life is shaped by and grounded upon facts about human beings, such as the need for food and shelter, but also the need for and interest in complex social relationships, aesthetic appreciation and entertainment, for example. To live one's entire life in a comparable state to Hare's interment would be to have a severely impoverished existence. Many people in the world do live such an impoverished existence, but unless the world has reached its real optimum population level, the opportunity to flourish will exist. The fact that people do not, in practice, have this opportunity raises important questions about social and political justice, which I will not address here.

Establishing the optimal population size

A further assumption concerns maintaining the optimal population level and preventing overpopulation. Throughout, the phrase 'birth rates' refers to the number of children born in a given year; 'mortality rates' refers to the number of people who die in a given year. Population growth is the result of birth rates that are higher than mortality rates for any specified period. The restriction of birth rates, either voluntarily or involuntarily, provides the usual means for controlling population growth. I will assume that reproductive rights may justifiably be restricted, at least to a limited extent, and that the threat of overpopulation is an acceptable justification for so doing. I also assume, however, that there is a minimum birth rate, that is, that a number of children will be born each year. Indeed, the minimum birth rate will determine the extent to which birth rates may justifiably be restricted. This minimum number of children, though not large, will be substantial. It should not be confused with the actual birth rate of a given population, which may be substantially greater than the minimum; it simply maintains that even with falling fertility rates in the West, there would be a minimum number of children born each year in a given population. What is more, it is desirable that at least this minimum number of children is born each year. As such, the claim that there is a minimum birth rate is both normative and descriptive, and I offer the following five reasons in support of this claim.

First, reproduction is a fundamental feature of our existence. Indeed, the Disposable Soma Theory's explanation for why we age rests on evolutionary reasons for why we reproduce. While this biological point is obvious, it is one that needs emphasising, particularly given how it shapes many of our values. Second, children are individually valuable for their own sake, not just as valuable additions to the community. Third, children need other children, that is, they require a community of other children and the culture that this community gives rise to, in order to flourish. The culture of childhood is as intrinsically valuable as most other cultures.

Fourth, Ansley Coale (1975, p. 40) maintains that a society which lacks a steady youth base, and is therefore on average older, will be more cautious and conservative. Indeed, Thomas Kuhn (1996, p. 90) notes that revolutionary changes in science are often brought about by the young. Kuhn also observes that changes are brought about by those who are new to a subject and consequently, an 'older' society might not be as conservative as Coale suggests, provided older generations change professions periodically. These are clearly contentious claims, but it is worth noting that if life spans increase as a result of an increasing ability to combat degenerative disease, the possibility of death by misadventure will take on greater significance and importance.

Finally, the young are essential to the economy, not least because of the extensive industries, such as education systems and youth entertainment, that respond to their specific needs. Moreover, economies rely upon the workforce to support welfare systems (Daniels, 2008, pp. 164–7). A stable youth base, the influx of new, young people into the economy, is necessary to sustain the workforce and support such a system. As life spans increase, along with a concomitant prolongation of health, the fact that new people to a given population are young may prove to be less important. Furthermore, as generations become older, the size of youth-based industries will reduce and industries aimed at older people will increase. Nevertheless, economic alterations reflecting the change in the average age of a population with significant increases in longevity will not remove the need for and desirability of an influx of some young people into society and its economy.

Some of these reasons I take to be obvious; others require further elaboration and support in order to be more convincing. Together, these five reasons suggest it is reasonable to believe that not only is it desirable for there to be a minimum birth rate but also that such a minimum rate can be expected. The reasons I propose for this minimum, particularly the first three, refer to fundamental features of human biology, values and needs. All five of the reasons also point to a broader, social understanding of human flourishing. The pursuit of the good life is one that takes place within a social setting because we are social animals (or Being-with-others). It follows from these reasons in support of the idea that there is a minimum birth rate that to reduce the birth rate so that it falls to below this minimum would involve curtailing basic human desires and values, which would in turn have a negative impact on the opportunity to flourish.

An asymmetry in the obligation not to have children

When a society is at its real optimum population level, the addition of an extra person will cause overpopulation, removing the opportunity to flourish for all members of society. The addition of an extra person at the real optimal population level will, therefore, involve bringing into existence a person whose life would not be worth living.[3] While most people agree that there is no obligation to have children whose lives would be worthwhile, there is an obligation against deliberately having children whose lives would not be worth living. It is important to emphasise that the idea of a minimum birth rate does not imply an obligation to have children who could flourish. The minimum birth rate concerns the influx of new people through birth into a given population and the desirability of this influx.

What constitutes deliberately having a child is open to dispute, and in this context would require that potential parents are aware that their society is at its real optimal population level, but I will assume that these difficulties can be overcome.[4] What is more problematic about bringing a person into existence is that the asymmetry in our obligations is difficult to justify. If whatever would make a person's life worthwhile fails to provide a reason for bringing them into existence, then whatever would make a person's life not worth living also fails to provide a reason against bringing a person into existence (McMahan, 2002, p. 300).

There have been many attempts to resolve this problem of asymmetry. Jan Narveson (1967) argues that welfare considerations of actions must be restricted to those people whom the act directly affects. What is more, we cannot benefit a person by bringing them into existence: to claim otherwise requires comparing their existing state with non-existence, which involves a referential failure. On the basis of these two claims, he justifies the asymmetry by arguing that bringing into existence a person whose life is not worth living involves making an actual person suffer. By failing to bring into existence a person whose life would be worthwhile there is no actual person who fails to benefit from having a worthwhile life. Both Narveson's claims, that welfare considerations must be person-restricting and that we cannot benefit a person by bringing them into existence, are questionable. For example, to avoid their child having a life that is not worth living, parents might delay conception by a number of years, but given the Time-Dependence Claim, a different child will be born, and so the reason for delaying conception may be impersonal rather than person-affecting (Parfit, 1984, pt IV).[5]

Although the asymmetry is difficult to justify, it is widely accepted, and I will assume that it is defendable. Whether I need such a defence is similarly disputable. For a non-consequentialist, the relationship between good-making properties and right-making properties is complex, and what is right need not be good (Tooley, 1998b, p. 117). The asymmetry argument relates to what we should and should not do on the basis of what is good or bad for the person who is brought into existence. While there appears to be symmetry between the properties that make it good or bad to bring an individual into existence, these properties may be asymmetrical in their right-making characteristics: what makes bringing a person into existence good might not make it right (pp. 117–18).[6]

Carter's critique of Parfit

In consequentialist population theory, the Total Principle aims to maximise the total quality of life, whatever makes life worthwhile, by maximising the number of people with worthwhile lives. An alternative to the Total Principle is the Average Principle, which aims to maximise the average quality of life of a given population. Parfit (1984, 1986) argues that both principles lead to the 'Repugnant Conclusion': a very large population where people's lives are only just worth living. In looking to avoid the Repugnant Conclusion, Parfit seeks a new Principle of Beneficence, one that would take into account society's obligations to future generations.

An increase in life spans creates a similar problem to that with which Parfit is concerned because it creates a conflict between extra lives and extra years, where an increase in longevity is an important feature of the quality of life. As a consequence, Parfit's discussion of population theory proves insightful for an assessment of the difficulties that increasing life spans will raise with regard to population growth. There have been many attempts at refuting Parfit's population problems and paradoxes, but I will focus on Alan Carter's critique (1999).[7] It is not the most recent attempt to refute Parfit, but it serves my purpose in demonstrating how we can conceive of a plurality of ways of flourishing. In so doing, it also provides a conceptual model for assessing the implications of increasing life spans on the population size.[8]

Parfit begins by contrasting two populations, *A* and *B*. *A* has a high average quality of life per individual and half the population of *B*. *B* has a lower average quality of life than *A*, but it has still more than half the average quality of life of *A*. In accordance with the Average Principle, *A* is better than *B* because lives in *A* go, on average, better. The Total Principle suggests *B* is better than *A* because the total sum of whatever makes life worthwhile is greater than in *A*. This is because the average in *B* is over half that of *A*, and there is twice the population. To emphasise the Total Principle, Parfit uses the analogy that two bottles that are each over half full contain more liquid than a full bottle (1986, p. 147). Such an analogy provides an initial reason to object to the Total Principle. Parfit is concerned here with a Principle of Beneficence that seeks to maximise the good (although, in general, Parfit is searching for a more subtle Principle of Beneficence). As Regan (1985, pp. 82–3) argues, this Principle of Beneficence is concerned only with maximising the liquid in the bottles, not the actual bottles: the Total Principle has little room for the individuals who contribute to the total quality of life.

It is because Parfit seeks to maximise the total sum of those things that make life worth living that the Repugnant Conclusion ensues. Just as the Total Principle claims that *B* is better than *A*, so *C* would be better than *B* if it had twice the number of people with an average quality of life over half that of *B*. Ever larger populations can continue to be imagined until the quality of life is barely worth living. The largest possible population will be one where the addition of another person will mean life will no longer be worth living. The Repugnant Conclusion exists because Parfit assumes the relationship between different population sizes is transitive (Temkin, 1997, pp. 300–11). Yet, as Carter (1999, pp. 304–8) argues, *B* might be preferred over *A*, considering a gain in numbers an acceptable trade-off for the loss of quality of life, but *C* is viewed as an unacceptable trade-off. The relationship between the different population sizes is non-transitive, and Parfit seems to neglect the possibility of unacceptable trade-offs between them.

To avoid the Repugnant Conclusion the Average Principle might be preferred, concluding that *A* is better than *B*. The Average Principle, however, also leads to some unfortunate conclusions. For instance, it would imply that it would be unacceptable to have a child with a high quality of life if the average quality of life of the population were higher (Carter, 1999, pp. 291–2; Narveson, 1973, pp. 80–1). Parfit also considers the Average Principle to lead to the Repugnant Conclusion. The Mere Addition Paradox demonstrates this.

Consider again populations *A* and *B*. In accordance with the Average Principle, *A* is considered better than *B*. Alongside this consider situation *A+*. Situation *A+* represents two populations, one that closely resembles *A*, while the other has the same number of people but a lower average quality of life, although life is still worthwhile. The inequality between these populations is natural, and neither population is aware of the other. Parfit does not consider situation *A+* to be any worse than *A*. To do so is to consider it better if the extra people of *A+* had never existed. The inequality of *A+* is unfortunate but not enough to consider the existence of the extra people in *A+* as undesirable. If *A+* is no worse than *A* the Average Principle fails because the combined populations of *A+* have a lower average quality of life than *A*. Parfit utilises the Total Principle – and the value of extra lives – to make this claim. Yet, *A* and *A+* are two very different situations with a non-transitive relationship. From an objective perspective, *A+* represents a world with twice the population and a large, natural inequality between societies compared to *A*. *A* might be preferred to *A+*, but this is not to claim that it is better or worse than *A+*. Such an evaluation cannot clearly be made.

Parfit then considers a situation where, owing to environmental changes, the average quality of life of both the populations in *A+* becomes more equal. Call this new situation *Divided B*. The combined average quality of life of *Divided B* is greater than the combined average of *A+*. The population most resembling *A* in *A+* has a lower average quality of life, while the smaller of the two populations in *A+* gains in average quality of life. Parfit considers *Divided B* to be a better situation than *A+* because of the higher combined average quality of life and the greater equality between the populations. Parfit's concern here is to emphasise the greater egalitarian nature of *Divided B*. To insist that *Divided B* is not better than *A*+ is to support the Elitist View: that the losses to those who were better off matter more than the gains to those who were worse off. While a preference for a more egalitarian society is admirable, the non-transitive relationship between populations of different sizes, quality of life and of equality suggests that there could exist reasons for preferring a society with greater inequality. As John Rawls (1971) argues, inequality might be acceptable if the whole community benefits from it.[9]

In a new situation, the two populations of *Divided B* discover each other and join. By chance, *Divided B* has the same number of people and the same average quality of life as *B*. The Mere Addition Paradox proposes the following three beliefs: that *A* is better than B; that *A*+ is no worse than *A*; and that *B*, which is the same as *Divided B*, is better than *A*+. As a consequence, the Mere Addition Paradox suggests that *B* both is worse and is no worse than *A*. This paradox can only be avoided by rejecting one of the three beliefs. Parfit proposes rejecting the belief that *B* is worse than *A*: to reject the belief that *B* is better than *A*+ involves supporting the Elitest View; and to reject the belief that *A*+ is no worse than *A* requires believing that it would be better if the extra people of *A*+ had never existed (1986, pp. 154–5). From this the Repugnant Conclusion ensues because the Mere Addition Paradox will imply that *C*, with twice the population and an average quality of life over half that of *B*, is no worse than *B* and so on until a population is reached where life is only just worth living. That it does so is only because Parfit exploits the Average, Total and Equality Principles at different stages of the argument. The Average Principle is exploited to suggest that *A* is better than *B*; the Total Principle is exploited to claim that situation *A*+ is no worse than *A*; and the Equality Principle is exploited to suggest that *Divided B* is a better situation than *A*+.

Parfit's singular application of particular values and principles continues with the Perfectionist Theory, which he proposes as a solution to the Repugnant Conclusion. The Repugnant Conclusion involves an unacceptable reduction of the quality of life, which will inevitably entail that a certain quality of experience will be lost, and quality experiences are often what make life go best. For instance, resources would not be available for a good performance of the music of W.A. Mozart, only a mediocre one. Parfit (1986, pp. 160–4) argues that when the quality of life is considered an understanding of experience is needed that is similar to J. S. Mill's conception of 'higher pleasures': those pleasures that are valuable because of their quality rather than a given quantity of them. As Carter (1999, p. 297) points out, Mill's higher pleasures could easily be factored into the Average Principle as a minimum standard, making Perfectionism redundant, but not solving the problems of the Average Principle. Moreover, Parfit again promotes one value, the quality of individuals' experiences, above all others. A good performance of the music of Mozart is valuable, but it might require considerable resources, such as a suitable auditorium. A mediocre performance of Mozart might be preferred if it releases resources that could be used to ease suffering, such as aiding the plight of refugees.[10]

Carter rightly argues that the reason why the Total Principle and the Average Principle (via the Mere Addition Paradox) lead to the Repugnant Conclusion is because they each seek to maximise a particular value. In so doing, they neglect the possibility of an acceptable range of trade-offs between the number of people and the quality of life. This range of trade-offs can be graphically represented by a simple indifference curve (Carter, 1999, pp. 306–7). Moreover, Carter criticises Parfit's population problems for focusing only on the values of the number of people, the quality of life and equality.[11] He argues that any decision concerning the optimum population size and the rejection of the Repugnant Conclusion must take into account these three values and the values of liberty and rights. The optimal population size allowing people to flourish will be determined by the acceptable trade-offs between these five values, which can be represented by a five-dimensional indifference curve. I will refer to this theory as the *Five Dimensions Theory*.

The indifference curve of the acceptable trade-offs will determine the optimal population size by stipulating when the addition of an extra person to the population requires an unacceptable trade-off. As I interpret Carter's theory, the indifference curve represents a range of states of affairs establishing the necessary conditions for human flourishing, where each acceptable state of affairs represents a different ideal of the good life. The optimal nominal population level, therefore, will be established by whichever position on the indifference curve a given society prefers. As the indifference curve establishes the necessary conditions for flourishing, no state of affairs on the curve will be better or worse than another except in terms of social preference. Carter (1999, p. 306) also observes that the indifference curve applies to all people, synchronically and diachronically, because the pursuit of the good life is grounded on basic human needs and interests. As such, the indifference curve represents and requires sustainable trade-offs establishing the conditions for flourishing for future generations.

The indifference curve of the Five Dimensions Theory provides a conceptual model for appreciating the pluralism of flourishing.

Fundamental to this pluralism are the acceptable trade-offs between our values. It is this framework that I will use to understand the consequences of increasing human life spans for the fair distribution of resources and ultimately the pursuit of the good life.

The problem of increasing longevity

The problem of overpopulation is one concerning the availability of sufficient resources so that every individual has an equal opportunity to flourish. Increases in life spans could threaten this availability in two interrelated ways. First, increases in longevity will involve a reduction of the mortality rate. Without a concomitant reduction in birth rates, this will lead to population growth. Second, any future increases in life spans will be the result of developments in biomedicines, which maintain or restore the quality of health by treating or preventing the ageing process and the diseases and disorders associated with it. This will entail that the demand for resources will increase for each individual. It is not obvious that the rise in demand will only be by the elderly. While treating the ageing process would lead to greater longevity, preventing the damage that leads to ageing may also be necessary, and this will involve medical interventions for the young. As a result, if a society is to accommodate increases in life spans, it may need to adopt a different position of the Five Dimensions Theory's indifference curve with a smaller optimal population level.

The consequent implications of longer life spans for population levels need not be problematic. It may be the case that it is possible to adopt a new state of affairs on the Five Dimensions Theory's indifference curve, or that a situation exists where there are sufficient resources such that society can accommodate both an increase in population size and the greater demand for resources by individuals without moving its position on the indifference curve. Nonetheless, with finite resources, increases in life spans could eventually threaten an unacceptable trade-off between our values. If we accept that the opportunity to flourish should be equally open to every individual, the trade-off required by increases in life spans will be between the quality of life and the number of people in society. I am assuming that if we can justify altering the quality of life or population levels, we can also justify their implications for rights and liberty.

The traditional means for altering population levels is to control birth rates and there may be some scope to reduce them. Nevertheless, I maintain that there is a minimum birth rate because of the fundamental needs and interests of human beings. In order to accommodate significant increases in life spans, it may be necessary to reduce birth rates to below the minimum level, which in turn will challenge these needs and interests. For example, many people who would want to have children, or simply a child, may be prevented from so doing. Fewer children will also change the culture surrounding childhood and the benefits this provides for children and society as a whole. The same basic requirements that shape the minimum birth rate ground our values, which contribute to the pursuit of the good life. As such, reducing the birth rate to below the minimum in order to allow for increases in longevity will involve an unacceptable trade-off between our values. In so doing, it will move society to a state of affairs that does not sit on the Five Dimensions Theory's indifference curve and so will not provide an opportunity for flourishing.[12] If there is a limit to the reduction of birth rates, there remain two alternatives. I provide a brief sketch of them here, and devote the remaining chapters to addressing the issues they raise.

The Five Dimensions Theory maintains that a trade-off can be made between the quality of life and the number of people in a given society. There are many goods that contribute to the quality of life and make life worth living. In order to accommodate increases in life spans, it will be possible to sacrifice some of these goods or their quality. For example, we might accept a mediocre performance of Mozart rather than an exemplary performance so as to allow the birth rate to remain above the minimum level while also permitting increases in longevity. This will involve moving along the Five Dimensions Theory's indifference curve to a state of affairs that promotes a different appreciation of the good life. There are limitations, however, to how much of the quality of life can be sacrificed: our activities and attachments have to be of sufficient worth to make flourishing possible. If too many of life's goods are sacrificed society would move to a state of affairs that does not rest on the Five Dimensions Theory's indifference curve. Longevity, however, is also a feature of the quality of life, and an obvious solution is to restrict increases in life spans. This will have the effect of trading some aspect of the quality of life to allow for an increase in the population and the greater demand by each individual for resources that longer life spans will involve.

The threat of real overpopulation implies that while it is good if a good life continues, it is good only if it continues for a limited period. This is no new claim, as I argued in Chapter 2 with regard to Williams's tedium of immortality argument. In both cases of the tedium of immortality and overpopulation, living longer risks undermining the pursuit of the good life. Although I argue that the risk the tedium of immortality poses will apply to all human beings, when it does so is specific to individuals. The tedium of immortality argument concerns the subjective effects of longer life spans on the individual pursuit of the good life. In contrast, overpopulation will undermine the objective conditions for the opportunity to flourish for all members of society. The threat of overpopulation suggests that while it is good for individuals if their life continues, it is good for present and future generations if life spans are limited.

The prevention of overpopulation by restricting increases in life spans will, however, require trade-offs in our values that might not be accommodated by the indifference curve of the Five Dimensions Theory. Just as abandoning the minimum birth rate requires reassessing the role of certain fundamental features about human beings in shaping the good life, so too will limiting longevity. It will require reassessing when and in what situations death is bad and the role and aims of medicine in enabling us to achieve and to continue living well. To restrict life spans will also challenge the view that human beings should equally have the opportunity to flourish for it requires that some people, namely the elderly, die so that others can continue maintaining and pursuing their activities and attachments.

There is a third option to restricting birth rates and mortality rates, and that is to abandon the idea of maintaining the real optimal population level. Rather than restrict access to resources, which limiting increases in life spans will entail, every individual could continue to have equal access to the resources they need in order to have the opportunity to flourish. Yet, with a greater number of people, each requiring more resources than is presently the case, continuing to allow equal access to resources may mean no individual has sufficient resources to provide them with the opportunity to flourish. The result would, therefore, also involve moving society to a state of affairs that does not sit on the Five Dimensions Theory's indifference curve.

5
Ending Lives

Longevity is an important aspect of the good life, but the discussion thus far identifies two reasons for limiting the prolongation of life. Unlike the tedium of immortality, which is likely to affect individuals only if increases in life spans are significant, the fair distribution of resources to ensure that every person has the opportunity to flourish is already a problem that modest increases in longevity may exacerbate. A third reason concerns the deterioration of the quality of an individual's life as they grow older. In such a situation, people may seek an end to their lives, either through their own or another's endeavours.

The value of longevity depends upon life being good, and what makes life good is maintaining and pursuing one's activities and attachments. As life spans increase, so social institutions will need to adapt to ensure that in an ageing society the opportunities exist for people to continue with what makes their life fulfilling. Fundamental to what makes life good is the quality of one's health. This is not to suggest that those with a low quality of health cannot live a meaningful and valuable existence, but to recognise that the quality of one's health will play a significant role in determining the content of one's life. The way in which society and social institutions respond to differences in the quality of health will also determine the opportunities people have with regard to their activities and attachments.

Increases in life spans will be the result of medicine that aims at restoring or maintaining good health in later life by preventing or treating the ageing process and the diseases and disorders associated with it. Francis Fukuyama (2002) warns that the unevenness in the development of medicines may make people live longer but without a concomitant prolongation of the quality of their health. The fact that ageing involves a number of complex systems, and that there remains much to learn about their functions and interactions, implies that there is and will be some degree of unevenness in the advances in medicine that prolongs life. How uneven this development will be, and more importantly, the consequences for the quality of health and life, remain to be seen. The way in which society responds to longer life spans without a concomitant continuation of the quality of health is important. It may be the case, for instance, that any decline in health can be compensated for by social changes designed to assist the elderly to maintain an active and communal life. Nevertheless, even where society is more responsive to and accommodating of the needs of the elderly, it may be that the quality of health for some is such that they would not want to continue living. As a result, increasing life spans may involve more suicides and requests for assisted suicide and euthanasia.

The tedium of immortality and the unevenness of technology point to subjective reasons for rejecting the prolongation of life. In both cases, others may be called upon to assist or act so as to end another's life. The fair distribution of resources raises different issues, focusing on how much effort and cost we should support in order to maintain and prolong life. As will become clear, the fair distribution of resources may also involve more suicides and requests for assisted suicide and euthanasia.

The fair innings argument

The pressure that increasing life spans will place on distributing resources fairly so that every individual has the opportunity to flourish means it might be necessary to restrict or prevent such increases. As argued in the previous chapter, there are limits to the reduction of the birth rate, but increases in life spans and the fair distribution of resources is not simply a matter of the number of people in a given population. Even in the case of modest increases in longevity the addition of quality years to people's lives will also require the use of more resources for each individual throughout the course of our lives. As such, one solution to resolving the distributive difficulties involved with increases in life spans is to limit them.

It is not my aim to consider all the arguments in favour of different types of distributive justice. I will consider the implications of increasing life spans, and in particular, limiting such increases on the 'fair innings' argument. My purpose in choosing this argument is to focus on the relationship between the notion of a normal life span and increasing longevity, and the implications of this for distributing resources, specifically healthcare. Increases in life spans will have a notable impact on the distribution of healthcare but, to be clear, it will affect the distribution of all resources.

The fair innings argument

The fair innings argument maintains that all members of society are entitled to sufficient healthcare to enable them to live in good health for the normal life span, that is, for a fair innings.[1] As was argued in Chapter 2, the notion of a normal life span informs our attitude to the misfortune of death. To live longer than the normal life span is to benefit from bonus years of life, but to die prior to it is to be deprived of a full life and the opportunities this presents.[2] On this basis, the argument maintains that healthcare resources should be directed towards enabling people to live for a fair innings and away from those who have lived beyond it. In so doing, the fair innings argument balances the demands for healthcare efficiency, which Alan Williams (1997) measures in terms of quality-adjusted-life-years (QALYs), the years of quality life that a medical intervention provides a patient, with demands for equity. The argument requires that every individual has sufficient healthcare resources to provide them with an equal opportunity for living in good health for the normal life span. A trade-off is made with the demand for equity to ensure that every individual has this opportunity (that it is efficient) because it denies those who live longer than the normal life span equal status with regard to healthcare.

One objection to the fair innings argument is that it is ageist. It deprives those people who live beyond the normal life span healthcare to ensure better health opportunities for those who have not yet lived for a fair innings. As was considered earlier, Tolstoy lived longer than the normal life span but his death would have been a misfortune for him so long as he wanted to continue living. John Harris (1985, p. 101) argues, the desire to continue living one's life is all that is necessary to have an equal claim on healthcare resources. He supports the use of the fair innings argument only for rare occasions when it is necessary to choose between two individuals, for example, rather than as a general means of resource allocation.

Against such criticism, supporters of the fair innings argument can claim that it applies equally to all people, but rather than consider the healthcare needs of individuals at a particular stage of their lives it focuses on their life-time healthcare needs. Hence, every individual has the equal opportunity to live for the normal life span.[3] Nonetheless, the fair innings argument accepts that such an approach entails that some people, those who live longer than the normal life span, will be treated unequally. The purpose of the fair innings argument is to justify this trade-off in equity to gain greater efficiency for every person for a certain time. What it means for people to be equal and treated equally, and whether sacrificing equity in the way the fair innings argument does is justifiable raises fundamental questions about the idea of human equality that I address in the following chapter. My aim here is to show that the fair innings argument cannot justify the trade-off in equity for efficiency that it supports.

A further problem for the fair innings argument, and which relates to the value of equality, is its central claim that there is a general entitlement to medicine. Daniels (2008, ch. 2), for example, relates the right to healthcare to the equality of opportunity, where the healthcare needs of all people must be met if they are to have the opportunity to pursue their life plans. I assume that such a right exists; my aim is to consider the implications for such a right that increasing life spans will have. I ground my assumption in part on the observation of the previous chapter that since flourishing is a basic human need we should ensure that the opportunity to flourish exist for all people, synchronically and diachronically. While this might imply a right to medicine, it does not lead directly to such a right. A further argument concerning social justice that normatively connects healthcare needs to rights is necessary. I assume such a normative connection exists, with my argument in the following chapter providing some implicit support for it.[4]

The length of a fair innings

What constitutes a fair innings is by no means clear. The normal life span might refer to average life expectancy at birth (or as Williams (1997) considers, quality-adjusted-life-expectancy (QALE) at birth). A problem with such an account is that, within a given population, different cohorts of people may have different life expectancies at birth. For example, the doubling of the life expectancy at birth during the twentieth century would have meant that, as cohorts born early in the century approached their average expected life span, new cohorts were being born whose average expected life span at birth would have been considerably greater. While it might seem reasonable and fair to extend the cultural, social, economic, political and environmental benefits that led to increases in life expectancy to those with lower average life expectancies at birth, the fair innings argument does not explain why this should be so.

It cannot explain why because there is no relationship between the statistical measurement of the normal life span and the idea of a fair innings. The fair innings argument is a prescriptive argument for the equal distribution of healthcare resources, as are all claims for equal treatment. Such prescriptions, however, must be grounded upon some feature or group of features of those beings that are to be considered equal. Average life expectancy at birth provides no such common feature. If the fair innings argument is to justify extending those benefits that give rise to longer average life expectancies at birth to those with shorter life expectancies it must fit within a broader moral framework or appeal to such a framework. In either case, it is this framework that justifies the fair distribution of healthcare and not the fair innings argument.

An alternative account of what constitutes a fair innings identifies it as the natural life span of human beings, that is, the biologically constrained length of life for which human beings can live. As such, the idea of a natural life span gives sense to the argument's claim of an entitlement to healthcare for the length of the normal life span. The normal life span, understood as the biological warranty period, grounds the entitlement to healthcare for the duration of the fair innings on a biological feature of human beings, in a way that is similar to how our medical needs ground the right to healthcare. The biological basis for the natural life span provides a common fact, that is, a universal and objective feature of human beings, thereby grounding the equality claims of the fair innings argument. The fair innings argument, defined in this way, would not need to appeal to an external account of equality in order to distribute social, cultural, economic and environmental gains to different cohorts of people.

The natural life span

To identify the normal life span as the natural life span, as the species-typical biological limit to human longevity, strengthens the fair innings argument. Nonetheless, the idea of a natural life span is disputable. Given the discussion thus far, to claim that the idea of a natural life span is disputable might appear contradictory. In Chapter 1, the Disposable Soma Theory was shown to be, of the theories considered, the most plausible explanation for what Carnes et al. (2003) refer to as the biological warranty period, that is, the species-typical life span. Even in a controlled environment, where external causes of death are excluded, human beings, along with most other animals, have a finite life span. The Disposable Soma Theory explains that the limit to human longevity is because we age, which is a result of the gradual accumulation of cellular damage. The biological warranty period, of approximately 85–95 years, reveals how much cellular damage a typical human body can withstand. As was argued, the advantage of the Disposable Soma Theory is that it can explain why there are differences in the species-typical life span of the various species but it also explains why individuals within a species can have different life spans. As a consequence, there would appear to be a natural life span by which to define the fair innings.

What makes the notion of a natural life span disputable is the role of human activities in determining how long we can live. Indeed, the difficulty with defining the normal life span in statistical terms is evidence of this. Life expectancy at birth can vary between birth cohorts because of differences in social, cultural, economic and environmental factors, where longer life expectancies reflect greater or more progressive forms of human intervention and developments. The prospect of increasing human longevity and further increases in life expectancy at birth beyond the so-called biological warranty period continue this process of human intervention in how long we can live, albeit more directly. This is not to claim that biology plays no role in determining the human life span, but it is how we respond with medicine, broadly conceived, to our biology that greatly influences how long we can live. The normal life span, understood as the limit to human longevity, will be defined by developments in and access to healthcare.

It might be countered that this denial of a natural span fails to distinguish sufficiently between the intrinsic and extrinsic restrictions on longevity. Such a counter argument, which I will refer to as the *Natural Life Span View*, maintains that human activities which focus on extrinsic limitations to longevity are enabling while future increases in life spans are enhancing. To control extrinsic factors, such as the environment, enables people to live for the duration of the natural life span; to alter an individual's biology so that they can live longer than the natural life span is to enhance them. The Natural Life Span View maintains that there is a natural, that is, a species-typical biological limit to human longevity. Human activities are enabling when they provide individuals with a greater opportunity of living for as long as the natural limit; they are enhancing when they interfere with this biological limit and make life spans longer. It should be noted that the Natural Life Span View does not imply that increasing longevity is impermissible, only that to increase life spans beyond the natural limit would be, descriptively at least, an enhancement.

Earlier, two accounts were given of the normal functioning of soma cells. The Natural Life Span View supports the teleological account, while the practical account, the idea that ageing is a disease, entails the denial that there is a natural life span. What gives rise to the view that ageing is a disease is the potential of biomedicine and the developments of gerontology, which suggest that it will be possible to prevent or retard the ageing process. This is not to deny that there is a natural impediment to living longer than the so-called biological warranty period. It is, however, to identify ageing as but one of the many obstacles to continuing to live. Human activities that the Natural Life Span View considers enabling also respond to these obstacles when they improve individuals' housing conditions and economic situation, protect against the environment, and address those impediments that are the focus of medicine. The aim of these enabling activities is not simply to prolong life but to restore or maintain a good quality of health, which in turn contributes to living well. Endeavours that prevent or treat the ageing process and the diseases and disorders associated with it, continue these activities. Future increases in life spans will not be the result of specific endeavours to increase longevity, but will be a consequence of medicine that aims at maintaining and restoring good health. The activities the Natural Life Span View considers to be enhancing have the same purpose and are a continuation of those it views as enabling.

To appreciate the rejection of the Natural Life Span View, consider a situation whereby increases in life spans are such that average life expectancy is 150 years. The Natural Life Span View maintains, and rightly, that if the biomedicine that produces such longevity were prohibited for those who reach the age of 90 and above, the average life expectancy would fall and no individual would be able to live to the age of 150. In contrast, if those activities that give rise to life expectancies of 90 were prohibited, average life spans would also fall, but some individuals might be able to live for as long as the previous average of 90 years. For example, average life expectancies in Ancient Greece were approximately 30 years, but some people, such as Sophocles, lived for 90 years. Even without the social, cultural, economic and environmental benefits that relatively recently have led to average life expectancies that are close to those of the alleged natural life span (the biological warranty period), it is not impossible for people to reach this age. Nonetheless, this fails to show that how long we can live does not depend upon human endeavours. For the Natural Life Span View to succeed, it must consider how long individuals, on average, would live without the benefit of any human activity. Such a life would be impossible, or at least very short, because to live from day to day requires that we satisfy our basic needs, such as those for food and shelter and in general to maintain our health. The activities that the Natural Life Span View regards as enhancing are simply more advanced means for prolonging life. It is for this reason that the idea of a natural life span is denied. How long we can live, and how long we will each live, is greatly dependent upon human activities, with ever-increasing life expectancies a result of greater progress in responding to and promoting our basic needs and interests.

Problems for the fair innings argument

The denial of a natural life span undermines the fair innings argument's justification for the redistribution of healthcare. The notion of a natural life span best defines what constitutes a normal life span for the fair innings argument because it grounds the entitlement to healthcare on a fundamental biological feature of human beings. For the fair innings argument, healthcare resources are distributed fairly only when every individual has the opportunity to live in good health for the normal span of life that is typical for a member of our species. Hence, only those people who have yet to live for as long as the normal span of years are entitled to the full range of healthcare.

In rejecting the notion of a natural life span, the normal life span, that is, how long people can live, becomes defined by human activities. More specifically, with regard to present life expectancies and future increases in longevity, how long we can and might reasonably expect to live depends upon advances in and access to medicine. Such an account of the normal life span has two important and related implications for the fair innings argument. First, the fair innings argument risks imposing a normal life span on people. A situation might exist where it is possible for people to live considerably longer than 90 years (the approximate length of the biological warranty period) because of advances in medicine that treat the ageing process. If the fair innings argument considers the normal life span to be approximately 90 years, it would deprive people of access to this medicine, thereby establishing a normal life span, rather than responding to one.[5] This may be necessary, as has been argued, in order to distribute resources fairly, and not simply healthcare, both for present and future generations. As has also been observed, if life is good, more life is better than less. To limit longevity brings into question such a view: it may be good for individuals if their fulfilling life continues but not for society.

Any restriction on life spans by limiting access to healthcare points to a second and related problem for the fair innings argument. The argument maintains that individuals are entitled to have sufficient healthcare so as to provide them with the opportunity to live for the normal span of life in good health. If the normal span of life, however, is largely determined by medicine, to which every individual is entitled, the fair innings argument cannot justify imposing a shorter life span by denying people access to medicine.

There may be a way to modify the fair innings argument so that it can accommodate or reject these problems, but it is not my aim to consider how this might be done. My purpose in discussing the fair innings argument has been to emphasise certain issues relating to the length of life, resource allocation, the nature and scope of medicine and the principle of equality. The scarcity of resources and the pressure that increasing life spans will add to distributing them fairly means it may be necessary to limit if not prevent greater longevity. The importance of the fair innings argument is that the trade-off between equity and efficiency that it proposes resembles that which curtailing increases in life spans involves. The fair innings argument attempts to justify denying the equal entitlement to healthcare for some people so that every individual has the opportunity to live in good health for a certain number of years, specifically the natural life span. In the absence of a natural life span, an alternative justification for the trade-off between equity and efficiency that leads to imposing a certain life span is necessary. It is a justification that must explain why the impartial demand that every individual should have the opportunity to flourish for a similar period of time can remove the entitlement of particular individuals to live in good health for as long as is it is possible for them to do so.

Limiting life spans

A limit to longevity may be one way of distributing resources fairly, but preventing increases in life spans raises a number of issues, not least whether such a restriction is justifiable. Of concern will also be how to achieve a limit to longevity. In assessing what options are available to society, I make a number of assumptions.[6] First, I assume that life spans in the developing world will continue to increase. If this were not the case, a policy restricting how long we could live would not be necessary.[7] Second, I assume that resources are sufficiently scarce to warrant such a policy. Whether this is the case is disputable. It may be possible to develop technology that can provide a more efficient use of resources, and greater tax or health insurance contributions might provide for more fairly distributed healthcare. This would allow for some increase in longevity, but as I proposed in the previous chapter, resources are finite and technology can produce as many problems as it solves. My concern is with what a policy limiting longevity might involve should it be required. Third, I take as a starting point the assumption that a policy limiting longevity is justified. My purpose in assessing what such a policy would involve is to question whether this is so, which I develop further in the following chapter.

In assessing a policy limiting longevity, I draw a distinction between establishing a normal life span and establishing a maximum life span. A socially determined normal life span is one that limits average life expectancies to a certain age, but where some people could and would live longer than the norm. In contrast, a maximum life span establishes a strict limit to how long people could live, such that no individual would, other things being equal, live longer than this maximum.

A duty to die

One way of establishing a normal or a maximum life span is to insist that those who reach the predetermined limit to longevity have a duty to die. John Hardwig (2000), for example, proposes that those people, specifically the elderly, have a duty to die when they become a burden for their families because of poor health. He does not deny that families have an obligation to care for their relatives when they become sick or aged. Nonetheless, he argues that the obligation to make sacrifices to support our families works in both directions, and those who are a permanent burden should sacrifice their lives for the benefit of those family members who support them.[8]

In a similar way, a duty to die might exist for those who reach the socially determined limit to longevity, requiring that they sacrifice themselves so that others may have the opportunity to live a life of good quality for the same number of years. A duty to die in this context places the responsibility for restricting life spans on those who reach the pre-determined normal or maximum life span. The duty might involve individuals refusing medical treatments that could prolong their lives, thereby establishing a normal life span; or it might involve directly ending one's life at a certain age, establishing a maximum life span. The duty to die assumes that suicide, the intentional ending of one's life, is morally permissible. I consider whether this is the case in more detail later in the chapter.

A duty to die offers one way of limiting longevity, but it suffers from the problem that, as Narveson (2000, p. 29) observes, we do not generally have a duty to die for the benefit of others. One reason why we do not is because of the belief that we are equal and should be treated equally: no one individual's life is more important than any other. To insist that there is a duty to die in order to ensure that present and future generations can flourish challenges a fundamental feature of morality. Yet, this is precisely the central issue involved in limiting longevity and which is demonstrated by the fair innings argument, that is, sacrificing the idea that we are equally valuable throughout our lives to achieve an equal opportunity for flourishing for a similar life span. If the trade-off is justifiable then appeals to equality will not be sufficient to dispute the idea of a duty to die.

An alternative objection to the idea that those who reach the imposed limit to longevity have a duty to die is the nature of duties. Narveson (2000), drawing on J. O. Urmson (1958), points to a defining characteristic of duties being that they are enforceable: sanctions can be brought against those who fail to perform their duties without reasonable justification. Given an entitlement of every individual to healthcare, in failing to perform their duty to die by refusing to give up the life-prolonging medicine that is available to them, an individual has wronged no one. The healthcare resources they receive are theirs by right and, as such, theirs to utilise or give away as they please. Only if an individual fails to utilise their healthcare and release it for others to use would those who could benefit from these resources have grounds for complaint.[9] A solution to this difficulty is to deny that those who reach the predetermined limit to longevity have access to healthcare, but this would remove the need for a duty to die, when it is understood as requiring the refusal of the medicine that is necessary to prolong life.

Although this feature of duties would seem to undermine or remove the need for a duty to die involving the refusal of healthcare, it may not prevent a duty to die involving killing oneself directly (by committing active suicide, which I clarify below) and so establishing a maximum life span. Nonetheless, a different aspect of duties does. When discussing the nature of supererogatory acts, Urmson (1958) observes that, in ordinary circumstances, a duty must be within the capacity of a person. This contrasts with supererogatory actions, such as heroic deeds, which involve acting in ways that are far beyond what we can normally expect of individuals. It is because duties are within the capacity of ordinary people that they are enforceable, unlike supererogatory actions, which are praiseworthy when undertaken, but are not blameworthy if they are not. The so-called duty to die involves a supererogatory act because it involves acting in a way that is contrary to normal human inclination.

The duty to die requires that people act in a way that is contrary to one's basic desires, those categorical desires that ground and provide the value and meaning for one's life, unless these desires are absent (in which case a person may well want to die). A duty to die requires that all people, irrespective of whether they want to continue living or not, end their lives when they reach the maximum life span. So long as the conditions exist for individuals to live fulfilling lives, it is reasonable to expect that carrying out the duty to die will be beyond the capacity of most people. Moreover, this will be the case both for a duty to die that involves people ending their lives directly and so establishing a maximum life span, and for one requiring the refusal of medicine, thereby establishing a normal life span. People who want to continue living will find it difficult, if not impossible, to refuse life-prolonging medicine. It would not seem feasible, therefore, to limit longevity by insisting that individuals take responsibility for so doing with a duty to die.

Restricting access to healthcare

As an analysis of the duty to die indicates, the most obvious way to limit longevity is to adopt a similar policy to that proposed by the fair innings argument. Once an individual reaches a pre-determined age, they would no longer have access to the medicines that prolong their life. An immediate problem with such a policy is that future increases in life spans will be a consequence of medicine that aims to restore or maintain the quality of health. The imposition of a limit to longevity means that in all likelihood the final years of life for those who live beyond this limit will be ones of a deteriorating quality of health. A similar situation follows from the fair innings argument. Despite its claims to justify depriving healthcare to those who live longer than the normal span of life, it is difficult to reconcile the argument's central premise of an entitlement to healthcare with allowing the deterioration of the quality of health, even for individuals who live longer than the normal life span. An entitlement to healthcare is grounded upon and is a response to our basic healthcare needs and interests. Medicine maintains and restores our health, understood as normal functioning. Yet, medicine, and the entitlement to it, is not simply a response to malfunctioning biology but to the suffering such abnormal functioning causes.

Williams (1997, p. 128) observes that it is unlikely that many people would interpret the fair innings argument as denying all forms of healthcare to those who live longer than the normal span of life. This may also be the case with imposing a normal life span, but it leaves open the question of the range of healthcare to which people who live longer than the age restriction would have access. As the purpose of establishing a normal life span is to control longevity, an obvious proposal is to deny people who reach the limit access to all medicines that can prolong life. This could involve a majority of healthcare options, with the important exception of palliative care. People who live longer than the limit might at least be entitled to healthcare that would ease any suffering they experience, so long as this does not make them live longer than they otherwise would have without it. An alternative solution is to prevent access to medicines that treat the ageing process directly. A problem with this alternative is the difficulty of distinguishing between those medicines that address the ageing process directly and those that treat the diseases and disorders associated with it because of a probable underlying connection between them. Successful treatments for age-related diseases, such as cancer, stroke and heart disease, may involve developing medicines that repair the cellular damage that gives rise to the ageing process.

The difficulty of determining which medicines should not be made available to those who reach the age restriction is further complicated by the claim that human activities determine how long we can live. The range of activities that contribute to the length of life extends beyond a narrow understanding of medicine as pharmaceuticals, for example. As the aim of establishing a normal life span is to limit the length of life, all activities that prolong life beyond the norm would need to be restricted. This might range from preventing people pursing healthy lifestyles to the logical extreme of prohibiting food. To deny people access to any resource that might prolong their lives would reduce how long people could live beyond the normal life span, and in so doing resembles the imposition of a maximum life span.

As undesirable as the imposition of a maximum life span in this way would be, it points to the fact that imposing a normal life span by curtailing access to healthcare after the age restriction may allow people to live some years beyond the limit. To curb how long people live beyond the age restriction, it may be necessary to ration healthcare prior to it. An inverse version of the QALY, for instance, might be used to determine whether a patient could have a specific treatment depending on whether it could make them live longer than the normal life span.[10] Such a policy would provide people with the opportunity to live for a certain number of years but not that these years be ones of good health, especially towards the end of the normal life span.

Limiting longevity, suicide and euthanasia

A third option for a policy restricting life spans is to kill those people who reach the socially determined limit and so establish a maximum life span. A policy of enforced euthanasia, as this option is sometimes referred to, has a precedent, according to Herodotus (1997, I.216).[11] A nomadic tribe called the Massagetæ sacrificially killed and ate the elderly members of their group; such was the value of this end that those who died of disease suffered the perceived misfortune of not being eaten. Despite its sacrificial nature, Robert Garland (1990, pp. 244–5) postulates that instituting an *ouros hêlikias* ('boundary marker of age') would have also served a practical purpose similar to that being considered here. A policy of enforced euthanasia would have a number of advantages over the above alternatives. First, it would, other things being equal, establish a maximum life span and without the need for individuals to overcome their categorical desires. Second, establishing a maximum life span could avoid the additional decline in health that might be expected with the imposition of a normal life span by restricting access to healthcare.

While it may have its benefits, killing people in order to establish a maximum life span violates a fundamental principle of morality. To refuse such a policy because of its implications for the prohibition on killing would not necessarily avoid a challenge to this principle. In rejecting a duty to die, the only alternative policy option available for establishing a limit to longevity is to refuse people access to medicine when they reach a certain age. Such a policy, however, may lead to a deteriorating quality of health for people who live longer than the age restriction, and possibly prior to it. A life of declining health, with no prospect that medicine will become available to improve one's situation, may lead some people to seek an end to their lives, by suicide, assisted suicide or euthanasia. What is more, the unevenness in technological advances may mean that any future increases in longevity that are possible may not be accompanied by good health, which in turn could also raise the issue of people seeking an end to their lives.

The right to life, understood negatively as providing protection against being killed, will prevent a policy that involves killing people in order to establish a maximum life span, but it may also prohibit suicide, assisted suicide and euthanasia.[12] In what follows, I will explain how and when suicide, assisted suicide and euthanasia are morally permissible. This argument will also demonstrate why a policy that establishes a maximum life span is wrong.

Before beginning, it may be useful to be clear about what is meant by suicide, assisted suicide and euthanasia.[13] Suicide is the intentional ending of one's life; assisted suicide occurs when one's suicide is aided in some way, such as being provided with the means to end one's life, where those who assist are aware of one's intentions. Euthanasia is the intentional ending of another's life as a means of relieving this person of some unbearable suffering; the motive for euthanasia is always what is best for the victim. Euthanasia is voluntary when the victim autonomously requests or consents to their life being ended; it is involuntary when no such permission is given; and it is non-voluntary when the victim is incapable of giving their consent to being killed. From this definition, it is wrong to identify the establishment of a maximum life span as enforced (involuntary) euthanasia: limiting longevity is not intended to be for the best for those who must die but for society. Throughout, my focus will be on justifying suicide and voluntary euthanasia. I assume that if these two activities are justifiable in certain circumstances, assisted suicide will also be justifiable in similar circumstances.[14]

All three forms of life-ending can be either active or passive. They are active when the means for ending life is the direct cause of death, for example, taking, providing or administering a fatal overdose of some drug. They are passive when the means for ending life is the indirect cause of death because it allows some underlying condition to progress unhindered. For example, euthanasia is passive when life-prolonging medicine is withdrawn from a patient with the intention that this should bring about the patient's death. The direct cause of the patient's death is their fatal condition and not the withdrawal of their medicine. In providing these brief definitions, I am obscuring some of the difficulties that can arise in distinguishing, for example, between passive assisted suicide and passive voluntary euthanasia, and between passive suicide and the non-suicidal request for the withdrawal of medical treatment. For the most part, these distinctions depend upon the intentions of those involved, but knowledge of the intentions for certain actions do not always explain matters. These difficulties will be clarified, when necessary, as the discussion progresses.

The sanctity of life

The sanctity of life doctrine, which traditionally informs Western laws and medical practices governing the ending of life, maintains that human life is of such special value that we commit a grave moral wrong when we intentionally end a human life.[15] The sanctity of life doctrine, as I interpret it, will not permit suicide, assisted suicide or euthanasia because they involve the intentional ending of a human life. There are three aspects of the sanctity of life that are informative for my assessment of the moral permissibility of killing a person.

First, the sanctity of life doctrine invokes religious connotations. For something to be sacred, according to the *Oxford English*

Dictionary, is for it to be an object 'secured by religious sentiment, reverence, sense of justice, or the like, against violation, infringement or encroachment' (5.a). The idea that life is sacred need not be religious, with Ronald Dworkin (1994), for example, providing a secular account of the sacred value of life, albeit one involving a number of difficulties.[16] The feature that I want to emphasise about the idea of the sanctity of life is that it is the value of human life that grounds the prohibition on killing. For some, such as John Finnis (1977) and David Oderberg (2000b), human life is a basic good of such special value that we cannot regard it as our own to do with as we please, and it is for this reason that suicide and euthanasia are deemed to be wrong. Throughout, I have argued that it is the content of life that provides it with its value. Thus the wrongness of killing, in part, relates to the badness of death, namely that it prevents the deceased from continuing to experience the fulfilment of their activities and attachments. This account of the value of life is crucial for the justification of suicide and euthanasia.

Second, claims about the value of life provide only one feature of the sanctity of life. With the exception of the most committed of pacifists, killing is viewed as morally permissible in some situations, specifically, self-defence and just warfare. The exceptions emphasise that while they are closely related, it is important to distinguish between the value of life and the conventions governing the ending of a life.[17] What is more, it does not follow that where killing is morally permissible the life of the victim lacks value or is no longer of a special value. For example, in the case of homicide in self-defence, it may be permissible to kill an unjust attacker because the conventions governing the ending of life permit such an action and not because the unjust attacker's life lacks value (Uniacke, 2004b, pp. 61–4). And it is the role of these conventions governing the ending of life that also explains why killing is wrong, as will become clear as my argument progresses.

Third, and related to the second aspect, the conventions of the sanctity of life doctrine governing killing state that it is wrong intentionally to end the life of a person. A moral difference is sometimes drawn between active euthanasia and passive euthanasia based on the idea that omissions – allowing a state of affairs to occur – may be less morally culpable than acting to bring about a state of affairs (Rachels, 1975; Singer 1993b, pp. 206–13, although both deny that there is a moral difference between acts and omissions). The sanctity of life doctrine as I define it may accept that there is a moral asymmetry in some cases between acts and omissions, but this is irrelevant in the case of active and passive euthanasia. Both intend the death of an individual and as such both violate the convention of the sanctity of life, which maintains that it is wrong intentionally to bring to an end a human life (Keown, 2002, p. 42).

As was observed above, intentions play a vital role in determining the moral permissibility of killing. It is by recourse to the Doctrine of Double Effect that the sanctity of life can coherently maintain that it is always wrong intentionally to kill and that ending a life can be morally permissible in just warfare and self-defence. The Doctrine of Double Effect permits actions that intend a morally good outcome but which have a foreseen but unintended morally bad consequence, where this consequence follows from and is proportionate to the intended outcome. For example, homicide in self-defence may be morally permissible only when the death of an attacker is proportionate to the threat they pose and is the foreseen but unintended consequence of the intended outcome of self-defence.[18]

The Doctrine of Double Effect will not only allow homicide in self-defence and just warfare, where killing is unintentional, but is also essential for the sanctity of life doctrine in permitting certain medical practices that will hasten the death of patients, such as the withdrawal of futile or burdensome medical treatments. The doctrine can serve to explain why the sanctity of life doctrine can permit ending a life in certain circumstances, while coherently prohibiting suicide, assisted suicide and euthanasia, but it is not without its difficulties. H. L. A. Hart (1967, pp. 122–5) and Philppa Foot (1967, pp. 23–4) point to the problem of 'conceptual closeness' between the intended outcome and its foreseen and unintended consequence, where it is difficult to draw a clear distinction between the two, such as with the withdrawal of burdensome life-prolonging medical treatment. A response to this difficulty is to consider whether the survival of the patient frustrates the outcome of the intended action. If it would not, then the death of the patient is not intended (Uniacke, 1994, pp. 106–7).

The right to life

One aspect of the sanctity of life doctrine concerns the value of human life while another concerns the conventions governing the ending of a human life. The right to life encapsulates these conventions. There are two features of the right to life that are crucial to my justification of suicide and euthanasia. The first concerns the relationship between rights and duties, while the second concerns the nature of rights and duties, particularly when they are in conflict.

The relationship between rights and duties

There are two types of relationship between rights and duties that are of relevance to this discussion, which, drawing on W. D. Ross's account (1930) of rights, I refer to as *Contract Rights* and *Correlative Rights*.

In the case of Contract Rights, the right to an object implies a conditional duty in relation to that object, and vice versa. For example, if *A* purchases an object from *B*, *A* has a right to the object, and this right implies a duty for *B* to provide it. What is more, *B* has a right to be paid for the purchased object and *A* has a duty to pay for it. If *B* fails in their duty to provide the object to *A*, *A*'s duty to pay for the object is cancelled, and *B*'s right to be paid for the object is also cancelled. If *A* refuses to pay for the object, *A*'s right to the object is cancelled, as is *B*'s duty to provide it. In contrast, the relationship between Correlative Rights and duties cannot be cancelled: the failure to fulfil one's duty does not remove another's right in relation to this duty. For instance, failure in one's duty to tell the truth does not cancel the right of others to be told the truth.[19]

Ross maintains that Correlative Rights are natural rights, where they describe the relationship that exists between moral rights and moral duties (p, 56). Natural rights, along with duties and obligations, are abstract nouns that serve as markers delineating appropriate moral behaviour between people. Rights identify those claims and powers that we have with respect to our moral values, which are grounded upon and reflect our fundamental needs and interests. Thus, the right to life conveys the conventions, or patterns of appropriate behaviour with regard to the value of human life.[20]

Conflicts between rights to life

As a natural right, one grounded upon basic facts about human beings, it is often maintained that the right to life is an absolute right that 'would remain in one's possession, fully effective as a ground for other people's duties to one, in all possible circumstances' (Feinberg, 1978, p. 97). This is difficult to maintain if we allow that killing is morally permissible in certain situations, as is accepted by the sanctity of life doctrine. The Doctrine of Double Effect permits homicide in self-defence when the death of the assailant is unintentional, but this raises the question of what happens to the assailant's right to life. As an absolute right, killing an assailant violates their right to life even when so doing is morally permissible.

In his assessment of the right to life, Joel Feinberg (1978) points to three methods for explaining what happens to the right to life in situations of permissible killing. Feinberg argues that the existence of exceptions to the prohibitions on killing means the right to life cannot be absolute in the sense defined (pp. 98, 104). He identifies three methods for resolving conflicts between two or more individuals' right to life, which purport to show that the right to life can be overridden while remaining absolute. Whether or not the right to life is absolute, or inalienable as Feinberg prefers (pp. 110–18), is not of concern here. What is of concern are the methods for resolving conflicts between rights that Feinberg outlines.[21]

One solution is to claim that there are situations where the right to life may be justifiably infringed, with the right either becoming ineffective or the deceased individual having forfeited the right to life because of their conduct (pp. 101–3). For this solution to succeed, an explanation of what constitutes a justifiable infringement is needed, but any explanation risks turning this solution into either of the alternatives. The explanation might specify the conditions for a justifiable infringement of the right, but this is the second solution; or it might be conditional upon moral judgements of the situation in which the right is infringed, but this constitutes the third solution.

The second solution is to provide a specification for the right to life, identifying all those circumstances in which killing, and hence violating a right to life, would be morally permissible (pp. 99–101). As a consequence, the specification of the right explains when it may be justifiably infringed. The account of the sanctity of life, as I have defined it, provides an example of the specification of the right to life: it specifies that it is wrong intentionally to end a human life. As it stands, however, this specification is too vague for it would permit any form of unintentional killing, thus allowing negligent behaviour. The exceptions to the prohibition on killing, namely homicide in self-defence and just warfare, are responses to the unjust conduct of those people whose right to life is justifiably infringed. The need to clarify the specification of the prohibition on killing in this way emphasises a potential flaw with this solution to conflicting rights. On the one hand, the specification of the right to life may be too vague, thereby allowing too many exceptions; on the other hand, if the exceptions to the prohibition become too specific, the right to life will not be able to accommodate new circumstances of permissible killing.

An obvious response to either difficulty is to modify the specification of the right to life to accommodate new cases or to restrict cases of permissible killing as they arise. So doing, however, points to a more fundamental, epistemic problem for the specification of rights, namely knowing when and why a certain specification of the right to life is justified. What is more, if it is possible to recognise when the specification of the right to life needs adapting, it brings into question the need for a specified right to life. Whether a violation of the right is permissible could be determined by a judgement about the particular situation in question, but this represents the third solution to resolving conflicts between rights.

The third solution to conflicts between rights discussed by Feinberg, the *prima facie* theory, addresses this epistemic problem (pp. 98–9). The *prima facie* right to life theory maintains that in normal situations the right to life is an actual right. In a situation of conflict, between two different rights to life, both rights are *prima facie* until it is decided which right is the more stringent. The more stringent right becomes an actual right and the less stringent right remains a *prima facie* right. The *prima facie* right to life remains present in all situations where it is at issue, but where it is not an actual right it might be perceived as having been justifiably infringed. Because the term *prima facie* suggests that rights are not real and incumbent upon those involved, the term *pro tanto* is sometimes used in its place to emphasise that rights (and duties) that do not become actual do not disappear.[22] By referring to the more stringent right in a given situation as actual further emphasises the perception that *prima facie* rights lack substance. Consequently, I will replace the term *prima facie* with *pro tanto*, and a *pro tanto* right or duty that becomes actual is the more salient, stringent or incumbent of the rights or duties that are in conflict.[23]

What determines which *pro tanto* right is more salient in a situation of conflicting rights is the use of reason. Ross, who first developed the *pro tanto* (*prima facie*) theory, is generally regarded as being an intuitionist. In fact, David Wiggins (1998, p. 272) notes Ross rarely uses the phrase intuition, with Ross instead claiming that we determine the most salient duty by a process of Aristotelian perception, which is the skill of practical reason.[24] It is through perception, where 'what I have to do is study the situation as fully as I can until I form the considered opinion (it is never more) that in the circumstances one of them [rights, in this case] is more incumbent than any other' (Ross, 1930, p. 19). Perception is the ability to bring reason to bear upon a particular situation and conclude what facts about the situation are morally salient. It is through reason that we can perceive what constitutes appropriate moral behaviour, for which natural rights, duties and obligations are markers. The ability to apply reason to perceive what constitutes right conduct and wrongdoing is fundamental to the theory of a natural law of justice.[25] In contrast to the specification of rights, a *pro tanto* account of rights provides general descriptions of rights and relies upon the ability to reason to determine when and whether killing is morally permissible. I discuss moral perception in more detail in the following chapter.

Suicide and euthanasia

The sanctity of life doctrine, as I portray it, will not permit suicide or euthanasia. Human life is of such special value that it is not our own to do with as we please. This value grounds the conventions concerning killing, which prohibits intentionally ending one's own or another's life. As a consequence, the right to life of the doctrine of the sanctity of life is, as Feinberg (1978, p. 110) describes it, a mandatory right: it is one that must be exercised as a duty to stay alive. In contrast, Feinberg argues that the right to life is discretionary and can be exercised negatively as a right to die, allowing the right's bearer to end their life without unjustifiable interference; or it may be waived, thereby freeing others to kill the right's bearer. While I concur with Feinberg that the right to life is discretionary and that suicide, in certain circumstances, is morally permissible, an additional argument is needed to justify voluntary euthanasia.

Suicide

Earlier, I noted Heidegger's response to the inevitability of our own deaths. An authentic response to this fact is to recognise that we are responsible for our own lives insofar as they are our lives to live. As such, we must choose to make our lives meaningful through the pursuit and maintenance of our activities and attachments. If we are free to choose the meaning of our lives, which makes life worth living, then we must also be free to choose to end our lives.

Indeed, how and when we die can contribute to the pattern and ultimately the meaning of our lives. Death is finitude: it marks the end of our lives. When we die might determine whether we have completed certain projects; or that we live beyond the completion of a project that defines us, which in so doing alters the meaning we sought for our lives. For example, someone who sought to shape their life by physical activity might consider the inactivity and frailty of their old age as undermining the significance, to themselves and to others, of their earlier life.[26] Just as when we die can affect the meaning of our lives, so can the manner of our deaths. One of the motivating reasons for Hardwig's claim (2000, pp. 122–3) of a duty to die is that many older people do not want to be a burden on their loved ones. While being a burden does not generate a duty to die, it might undermine the integrity of an individual's conviction that they should lead a largely self-supporting life. To end one's life, therefore, may be a means for maintaining the integrity of the personal beliefs and values that made one's life fulfilling.

Someone who supports the sanctity of life doctrine might agree about what makes life meaningful but deny that it follows from this that we can choose to end our own lives. As was also argued earlier, our freedom to choose the content of our lives is limited. We cannot choose our dispositions or the culture into which we are born. For the doctrine of the sanctity of life, a further constraint on our freedom is the special value of life, which makes suicide wrong. This constraint could prevent people from maintaining the integrity of their convictions, but in so doing it represents another aspect of the fragility of life. How and when we die is not always within our control, and this fact influences the values of our activities and attachments.

The difference in outlook concerning the moral permissibility of suicide emphasises a distinction between what Dworkin (1994) identifies as conservative and liberal approaches to the value of life. The doctrine of the sanctity of life represents a conservative approach to the value of life. For the liberal view taken here, what matters about life and makes it valuable is its content. This is not to deny that the human body is without value but to claim that what makes human life worth living are our activities and attachments. The earlier discussion of why death is a misfortune for the deceased emphasises why this is so.

Central to the dispute between the liberal and conservative approaches to the value of life is what limitations exist on our freedom. Both approaches accept that such limitations exist; nothing of the argument justifying the moral permissibility of suicide undermines my earlier claims about the constraints on maintaining and pursuing the content our lives. Such constraints, specifically our duties and obligations towards others, will limit the occasions when suicide is morally permissible. It would be impermissible, for example, to commit suicide as a means of avoiding punishment or to escape one's responsibilities (Aristotle, 1985, 1116a13). What makes suicide wrong in such instances is not the ending of a human life but the failure to fulfil one's responsibilities and the harm that might arise from such an action for others.

My concern, however, is with situations where an individual seeks to end their life because their activities and attachments no longer make it worth living. The reason for this may be health-related. An individual might seek to end their life because the suffering from a prolonged illness greatly outweighs and prevents them from experiencing the fulfilment of their activities and attachments; or a health condition may be such that while it does not cause physical pain it prevents a person pursuing a sufficient range of their activities and attachments for them to consider their life worth living. The reason for committing suicide might not be related to health, such as with the tedium of immortality, in which case an individual's activities and attachments no longer provide a reason for living. It is important to emphasise two points about these reasons for ending one's life. First, the reasons are subjective: individuals will disagree about the degree of suffering they are prepared to withstand or whether the fulfilment of their activities and attachments does provide them with sufficient reason to continue living. Second, countervailing reasons, such as having responsibilities to and for others, may act against these reasons. As with conflicts between rights and duties, whether suicide is morally permissible requires bringing reason to bear upon the facts of a particular situation.

Euthanasia

The argument in favour of the liberal approach to the value of life involves the freedom to choose, other things being equal, to end one's life, but it does not follow from this that it must be a result of one's own actions. The moral permissibility of suicide implies that the right to life is discretionary and we can waive our claims against others not to kill us. Yet, the right to life is a natural right and waiving it will not cancel the duty on others against killing. The right to life is a marker for one aspect of a particular form of appropriate moral behaviour between people; the duty not to kill is the correlate of this marker in relation to this behaviour. If voluntary euthanasia is to be regarded as morally permissible, it is necessary to show how and when the duty against killing can be overcome.

The solution to the problem of voluntary euthanasia is, I propose, that the duty against killing can be less stringent than an alternative duty in certain circumstances. What distinguishes a case of voluntary euthanasia from a case of malicious killing, for example, is the fact that voluntary euthanasia is by definition performed for benevolent reasons: the killer acts in order to relieve the victim of their suffering. Voluntary euthanasia, therefore, involves a conflict of duties between on the one hand, a duty of beneficence, specifically the duty of charity, and on the other hand, a duty of non-maleficence, specifically the duty not to kill.[27]

It is normally the case that duties of non-maleficence are more stringent than those of duties of beneficence because the former are perfect duties while the latter are imperfect duties. A perfect duty is one that can be fulfilled in only one way, in contrast to an imperfect duty where we can exercise discretion in how we fulfil it.[28] There is only one way in which we can fulfil the duty not to kill, which is, other things being equal, one we must exercise at all times, but there are a number of ways and different times when we can fulfil our duty to relieve suffering. The greater stringency of perfect duties means that when they conflict with imperfect duties they normally take precedence. Nevertheless, there are counterexamples which show that this is not always the case. For example, if someone is unjustly attacked, a third person may, in certain circumstances, intervene to defend the victim by killing the attacker. In this case, the reason why the duty of beneficence is more salient than the duty against killing is because of the injustice of the attack.

As with the moral permissibility of suicide and the conflicting duties in the example of defending a third person, reason must be brought to bear on the facts of a situation in order to judge which duty is the more salient. Two features about voluntary euthanasia are significant in determining its moral permissibility. First, the subject of voluntary euthanasia waives their right to life and in so doing signifies both that they want to die and that they want to be killed. Second, voluntary euthanasia is an act of mercy and is performed for the good of the one who dies. The individual who performs voluntary euthanasia must do so in response to their view of the value of life of the one who dies, but also in response to the subject's view of the value of their life. The fact that voluntary euthanasia involves fulfilling an imperfect duty is important because, for many people, euthanasia contradicts their deeply felt convictions about killing another person. There are many ways in which to be charitable, and so refusing to perform euthanasia will not necessarily involve failing to fulfil one's duty.

As I noted above, seeking an end to one's life is not straightforwardly about suffering, but the effect that the causes of this suffering and its consequences have on one's ability to pursue the activities and attachments that provide one's life with value. It is questionable, however, whether voluntary euthanasia would be morally permissible for the tedium of immortality. The tedium of immortality arises because the fulfilment of one's activities and attachments no longer provides a reason for continuing to live and not because they cannot be fulfilled. This implies that someone experiencing the tedium of immortality would, other things being equal, be capable of committing suicide or seeking assistance with their suicide. For such a person to request to be killed raises doubts about the sincerity of and the motives for their request.

The argument for the moral permissibility of voluntary euthanasia will not permit involuntary euthanasia because the subject does not waive their right to life.[29] Whether non-voluntary euthanasia is morally permissible is more difficult to resolve. In the absence of the ability to request or refuse euthanasia, the moral permissibility of euthanasia will depend upon third-party perceptions of the value of the subject's life. An advantage of the sanctity of life doctrine is that there is only one valid perception of the value of the subject's life, and this will prohibit euthanasia. The liberal approach must determine whether an individual's activities and attachments are sufficient to be meaningful to the subject, which involves the difficulties of knowing what the subject would find meaningful (if they had not or could not have determined this with an advance directive, for example) and agreeing about this. In some cases it is obvious that an individual's life is meaningless to them, such as that of a person in a persistent vegetative state. In such cases it is futile to keep their body alive. The moral permissibility of non-voluntary euthanasia is more difficult to determine in other cases, such as severe dementia, and each case will need to be judged on its own merits, taking into account what is known about what the subject would want in such a situation.

There also remains the question of what happens to the subject's right to life in cases of non-voluntary euthanasia. Voluntary euthanasia is permissible in part because the subject can waive their right to life; if the subject is incapable of waiving their right to life, it is reasonable to assume that it remains in place and fully effective. The argument in favour of voluntary euthanasia aims to explain why and when the duty against killing is not incumbent upon the one who performs euthanasia. In a similar way, any argument in favour of non-voluntary euthanasia must be sufficient to show why the right to life of the subject is not salient in the moral assessment of what may be done.

The argument I provide in support of the moral permissibility of voluntary euthanasia does not undermine the right to life as a protection against a policy imposing a maximum life span. For the most part, those people who live for as long as an imposed maximum life span would not waive their right to life because, other things being equal, their lives are meaningful and valuable to them. The discussion of euthanasia also maintains that to establish a maximum life span by killing people involves violating the duty against killing: it involves inappropriate moral behaviour by those who kill and those who sanction killing.

In contrast to establishing a maximum life span, a normal life span is established by denying those people who live longer than a certain age access to the broad range of healthcare. As a result, those people who live longer than the imposed limit may experience a decline in the quality of their health. For some people, but by no means all, the decline in the quality of health that could follow from living beyond the imposed normal life span may be such that they would seek an end to their lives. To what extent the quality of life would deteriorate will depend upon how great are the restrictions on access to healthcare and other resources. What the above argument shows is that where the decline in the quality of health and life is such that a person no longer wants to continue living, it is morally permissible for them to end their life by whichever method, be it suicide, assisted suicide or voluntary euthanasia, is appropriate.

6
Partiality and Equality

The threat of overpopulation and the specific pressures that increasing life spans will place on the fair distribution of healthcare resources means it may be necessary to restrict how long people can live. This is not to deny that there may be alternatives to limiting longevity; my concern is to consider some of the ethical implications of so doing. The analysis of the fair innings argument points to one such issue, namely the potential need to make a trade-off between demands for equality and the efficient use of healthcare resources. The trade-off rests on the claim that a fair distribution of healthcare resources (although the argument also applies to all resources) would involve restricting how long we can live to ensure that every individual, presently and in the future, has the opportunity to flourish for a certain number of years.

Whether this would be fair is questionable. The trade-off requires a fundamental alteration to the notion that human beings are equal in some basic way. This idea has been a feature throughout my discussion, be it explicitly in the consideration of the interests of animals and the problem of overpopulation, or more implicitly, where the natural right to life applies equally to all those who possess it.[1] In order to appreciate the nature of this trade-off, and what the justification for it must achieve, it is necessary to explore the idea of equality.

The principle of equality

An example of how the idea that people are equal features in my argument can be found with the initial objection to the notion of a duty to die. If every individual's life is of equal value, there can be no reason why one individual must die for the sake of another: no one individual is more valuable than another. I will refer to the claim that people's lives are of equal value as the *Principle of Equality*. As it stands, the Principle of Equality lacks any normative implications. If two objects are of equal value, to show a prejudice in favour of one of the objects is not necessarily wrong. For the Principle of Equality to lead to the equal treatment of individuals it must be supported by a further claim, that individuals should be treated according to the value of their lives. I refer to this additional claim as the *Normative Claim*. The prescriptive claim for equality must be grounded upon some fact or facts about people, which for the Principle of Equality is the claim that all human lives are equally valuable.

In his discussion of supererogation, Joseph Raz (1975) argues that a tension exists between the value of individual autonomy and the values of human life and human equality that give rise to the welfare of society. The welfare of society and the value of life are related by the claim that the activities and attachments that people maintain and pursue provide their lives with its value. A flourishing society is one where people are able to have a wide range of activities and attachments that satisfy their fundamental needs and interests, and which also allow people to provide (albeit largely unconsciously) their lives with a distinctive, individual pattern. Furthermore, individuals' lives can only be distinctive if they are autonomous and free to choose the activities and attachments they maintain and pursue. Yet, as has been observed, there are constraints upon this autonomy, some of which are intrinsic to the welfare of society. Central to these constraints is the Principle of Equality. The claim that people's lives are equally valuable implies that the maintenance and pursuit of activities and attachments by some people should not be to the detriment of others in society. Consequently, there is a tension between an individual's autonomy to live their life distinctively, and the promotion of the welfare of society, where everyone has the opportunity to maintain and pursue their activities and attachments. Raz (1975, p. 168) concludes that people are allowed to pursue the content of their lives unless so doing offends against fundamental moral requirements, where the Principle of Equality is a fundamental moral requirement (although Raz does not explicitly state this). This leads to the following statement:

(1) A person is justified in pursuing their activities and attachments except when so doing conflicts with the fundamental moral requirements.

Raz argues that supererogation is an exclusionary permission, one that permits people to avoid performing some action for the benefit of society, if that action might require sacrificing their fundamental concerns (p. 167). A person's fundamental concerns are the maintenance and pursuit of their activities and attachments, which fulfil their basic requirements (their fundamental needs and interests). In voluntarily choosing to risk their fundamental concerns in order to promote the welfare of society a person performs a supererogatory act. The Principle of Equality implies that a person is under no obligation to, nor can some group legitimately demand that they sacrifice their fundamental concerns to maintain or improve the welfare of society.

The need to limit how long we can live in order to distribute healthcare fairly involves a trade-off that redefines the Principle of Equality. Given the grounds for the Principle of Equality and the need to limit longevity, the redefinition would take something like the following form:

(2) The value of life declines after a certain age, where every individual's life is the same for any given age.

Statement (2) would replace the Principle of Equality as a fundamental moral requirement, which along with the Normative Claim would make it permissible to limit longevity by restricting access to healthcare. The value of life declines after a certain age, rather than falls immediately to some lower value in order to take account of the varying policy options that might establish a normal life span. With statement (2), every individual remains entitled to healthcare, but the Normative Claim implies that it is permissible to treat healthcare needs differently. Thus, those who live longer than the normal life span may still be entitled to palliative care, while those who are younger than the age at which the value of life declines are entitled to the full range of healthcare options.

Statement (2) explains roughly how the fundamental moral requirements would need to change in order to limit longevity, but it provides no justification for such a policy. One such justification is Daniels's Prudential Life Span Account (2008, pp. 171–81). The fair distribution of healthcare may be viewed as a conflict between the young and old. As society ages, because of declining birth rates and a reduction in mortality rates for all age groups, the profile of its members' needs alters. The elderly will require more welfare support because of the healthcare implications of ageing and a longer retirement period. In most Western societies, the burden of meeting this increase in needs will fall on the young (pp. 161–71) [2]

Daniels argues that rather than view the problem of resource distribution as one between generations, it should be viewed as one across a life: each generation represents a different stage of our lives (p. 172). With insufficient resources to cover our needs throughout our life span, a decision must be made about what stages in life it would be acceptable not to have these needs met (p. 173). The rationing of healthcare early in life might allow people to live longer than the normal life span in good health but reduce the chances of living for the full life span (p. 178). The rationing of healthcare later in life increases the likelihood of living for the normal life span in good health, which might also have the effect of creating more opportunities for maintaining and pursing one's activities and attachments. When a choice exists between age rationing – favouring the young – and a lottery allocating healthcare, where both will achieve the same life span, prudent healthcare planners will choose age rationing, which offers a greater chance of living for the full length of the life span in good health (pp. 178–9). In making this decision, the Prudential Life Span Account makes two caveats: first, those who decide the allocation of resources cannot know how old they are; second, they must be prepared to live with the consequences of their decision (p. 174).

Daniels recognises a number of difficulties with the Prudential Life Span Account. It is prudent when the only alternative is a lottery for allocating healthcare, but there may be other alternatives that are more prudent. What is prudent will also depend upon the range of resources that are open to decision-makers, although Daniels argues that the limitations on available resources do not have as much effect on prudence as is sometimes claimed (p. 32). My argument that human activities greatly determine how long we can live also brings into doubt what constitutes a prudent decision in the allocation of healthcare. Knowledge that there is no natural limit to how long we can live might make age rationing, and thus the shortening of how long we can live, seem imprudent. Daniels concedes that there will be disagreements about the moral permissibility of age rationing (p. 181). In effect, some people will object to statement (2) as a replacement for the Principle of Equality. Indeed, he considers the Prudential Life Span Account to be morally permissible in only relatively few circumstances, when resources are scarce, but my reason for considering statement (2) is the presumption of such scarcity.

The Prudential Life Span Account provides a justification for statement (2) but it is a limited and to some extent flawed one. In particular, it fails to provide a convincing argument against the Principle of Equality. There may be other justifications for statement (2), but in order to understand what they must achieve and why the Principle of Equality is for many a basic moral requirement, it is necessary to consider how people value each other.

Establishing equality

The claim that people are equal and should be treated as such has deep historical and psychological resonance.[3] Yet, as Williams (1962, p. 230) observes, the idea of equality, and the normative assertion that attaches to it, are difficult to explain. Human beings differ in a variety of ways such that it can seem impossible to find a feature that is common to us all. Should such a fact be found, such as Singer's claim that we have in common certain interests, it does not follow that this entails that we should be treated equally. There are many situations where an appropriate response to our interests would involve the unequal treatment of people.

To establish and justify the ideal of equality, specifically the Principle of Equality, Nagel (1991, pp. 10–20) argues that it is necessary to begin from the subjective perspective.[4] The value of an object is linked to the specialness of that object to a particular person. The objects people find special are those that matter to them in a way that other objects do not. One object that most people find special above all other objects is their own lives. As was noted in the discussion of the misfortune of death, Nagel also proposes that we view our world from both a subjective and an objective perspective. From the objective, impersonal perspective we can perceive that everyone views their lives as being special, which leads Nagel to claim that the 'basic insight that appears from the impersonal stand point is that everyone's life matters, and no one is more important than anyone else' (p. 11). Nagel goes further and maintains that if lives matter, they do so 'hugely' (p. 19). Consequently, he proposes, not only are people's lives hugely valuable, but they are equally valuable.

The impersonal perspective, John Cottingham (1997, p. 2) notes, resembles that of Adam Smith's and William Godwin's 'impartial spectator', the Ideal Observer discussed in relation to speciesism. While an impartial observer could view people as considering their lives to be special, it does not follow that people consider their lives to be equally special, or that the observer perceives them to be so. For instance, the Arrogant Egoist might perceive their life to be exceptionally special, while the Suicidal Cynic might consider their life to be quite dull and mundane. The notion of an impartial observer has similarities with that of Rawls's account (1971) of the Original Position behind a Veil of Ignorance, from which position people are unaware of features about their lives that might prejudice their decisions in developing moral principles. The Prudential Life Span Account relies upon a similar idea in justifying age rationing. Rawls maintains that in such impartial circumstances people would regard each other as equal. Yet no such assumption can be made. Alternative arrangements can reasonably be imagined in such situations that do not entail equality (Cottingham, 1997, pp. 4–5). Indeed, it is not obvious that the idea of equality would be a rational response to such a situation of uncertainty (Nagel, 1978, p. 121).

While an impartial observer could view people as leading equally valuable lives, an impartial perspective is, as was noted earlier, not one that any person can occupy. Furthermore, the impartialist places unjustifiable weight on equality when they suggest that if all people view their lives as special, people's lives are equally special. Any evaluation of a person's life will require a consideration of a wide range of values, both moral and non-moral, not all of which will be given the same weight (Berlin, 1956, pp. 319–20). For instance, Cottingham (1997, pp. 3–4) contrasts the value of a 'lager lout', whose main objectives in life are consuming copious quantities of alcohol and being disruptive, with that of a brilliant scientist. The impartial observer, Cottingham argues, would place a greater value on the life of the scientist than on the life of the lager lout because the activities and attachments of the scientist exhibit qualities that are valued in society above those pursued by the lager lout (p. 3).

In judging the lives of a brilliant scientist and a lager lout to be of unequal value, the alleged impartial observer assesses each individual's character. An individual's character is the expression of the range of moral and non-moral virtues that they possess, where the virtues are dispositions to undertake, and in an appropriate way, those activities and attachments that promote the good life. The content of a person's life reflects but also shapes their character, the type of person they are. If the brilliant scientist is of good character it is in part because the pursuit of science is valued, not only because of the contribution scientific achievements make to society, but also because scientific activity and achievement contributes greatly to the good life. The brilliant scientist will also be of good character because they are caring and generous to their friends and family, polite and sociable to people they do not know, and pursue a range of socially and culturally valuable activities. Contrast this with the life and character of the lager lout. Their socially disruptive activities and their liking of cheap, high-strength lager are not valued in society because they do not involve activities and attachments that promote individual flourishing, and which also disrupt the endeavours by others to lead a good life.[5]

It follows that a person of good character has a more valuable and fulfilled life because they are more likely to lead a good life than one whose character is less good. This conclusion gives rise to the following statement:

(3) The value of a person's life is assessed according to their character.

As the contrast between the brilliant scientist and the lager lout shows, people's characters differ, as do the values of the countless activities and attachments that they pursue and maintain. As a consequence, statement (3) conflicts with the Principle of Equality because from the impersonal perspective we will view people as being of unequal value in accordance with their character.

While statement (3) contradicts the Principle of Equality it would appear to complement statement (1), which asserts that a person is justified in pursuing their activities and attachments, because it is these activities and attachments that represent and shape their character. Indeed, statement (1) places limits on those virtues that contribute to a person's character by requiring that they do not conflict with the fundamental moral requirements. Furthermore, statement (3) might indicate that the restrictions of statement (1) are in some way morally defining because the assessment of character is a moral evaluation. It is for this reason that the assessment of a person's character from the impersonal perspective is not impartial: we cannot step outside of morality and the basic human pursuit of the good life.

Yet, statement (3) raises worrying concerns about elitism in any theory that promotes the value of a person's life according to their activities and attachments. Cottingham's argument, for instance, that the lager lout's life is less valuable than that of a brilliant scientist, could merely demonstrate a prejudice of Cottingham's in favour of certain activities demonstrated by the brilliant scientist and not by the lager lout. More importantly, the Principle of Equality, which statement (3) contradicts, is a fundamental moral requirement, where such requirements delimit, restrict and define what activities and attachments should be pursued in order to live well, and consequently what type of character is socially and morally desirable. For example, the sociable behaviour of the brilliant scientist is preferred to the loutish behaviour of the lager lout because the virtue of friendship accords with the fundamental moral requirements. The relationship between the virtues of character and the fundamental moral requirements is closer than this account suggests, as I will argue below.

Partialism

An impartialist perspective on our lives and specifically on our values would seem to offer the ideal support for the idea of equality and its normative implications. Every individual views their life to be special, which implies that every individual is equally valuable and should be treated as such. The above examples concerning character and the earlier doubts about the possibility of an impartial perspective raise doubts about the credibility of impartialism as the best way to understand our relationships with each other. The alternative, as the criticism of the Ideal Observer implies, is to adopt a partialist perspective.

Self-concern and moral considerations

Partialism is the idea that not only do we demonstrate favouritism towards our own activities and to certain other people but that we are justified in so doing. As such, partialism would seem to provide no scope for the idea of equality. Indeed, partialism obviously contradicts the idea of equality, specifically the Principle of Equality, because it justifies valuing and treating people differently. This need not mean that partialism cannot support the notions of equal rights or equality of respect. As a disposition and an interconnecting set of preferential relationships, partialism might, it is sometimes argued, provide the basis of morality (for example, Cottingham, 1997). This last claim, I will argue, cannot be a feature of a tenable account of partialism.

As an example of the difference between a partialist and an impartialist approach to ethics, consider a situation where a person must choose to save either their sibling, the lager lout, or a stranger, the brilliant scientist, from a burning house, but they cannot save both. Partialism maintains that the rescuer is justified in preferring to save their sibling because of their familial relationship. Indeed, it might be expected that they do so, and could demonstrate a flaw in the person's character if they were to choose the scientist (a point I clarify below).

In contrast, a partialist would be critical of the response of impartialist ethical theories, such as utilitarianism and deonticism, to such a dilemma. A partialist will accuse a utilitarian of considering only the utility value of the lager lout and scientist, where the utilitarian is expected to save the brilliant scientist because of their higher utility value. A deonticist will be accused of maintaining that there is an equal obligation to save either person. Yet the partialist's accusation requires very rigidly and narrowly defined conceptions of utilitarianism and deonticism. That the lager lout is the rescuer's sibling has a utility value, which the utilitarian would consider when making their calculation. Likewise, the deonticist might consider the obligation a person has towards their sibling as morally salient in this case. If impartialist theories, such as utilitarianism and deonticism, are to be plausible, they must take these considerations into account, and if they can do so partialism will not distinguish itself from them.

In response, the partialist will object that impartialist theories only allow the rescuer to choose their sibling when it is morally beneficial to do so, because it increases welfare or fulfils an incumbent duty. Impartialism will not allow us to choose in favour of our self-interests when there are other, morally beneficial alternatives (Wolf, 1982, p. 422).[6] The objection the partialist raises against impartialism is encapsulated by the failures of utilitarianism and deonticism to account for supererogation. An example of supererogation is that of a soldier who throws himself onto a grenade in order to protect his comrades. In so doing, the soldier considers this to be the right thing for him to do in that particular situation. What identifies his act as supererogatory is that were the soldier to decide not to sacrifice his life for his comrades, even though he considered this to be the most morally beneficial course of action, he would not be acting wrongly (Urmson, 1958, p. 63). Supererogation describes situations where we can place our self-concerns before those of moral considerations but choose not to do so. A supererogatory act is one that goes beyond the demands of duty.

The partialist's criticism of utilitarianism and deonticism is that they cannot, as impartialist ethical theories, make sense of supererogation. If a utilitarian decides that the right course of action is for the soldier to throw himself onto the grenade, it is because, taking account of the loss of the soldier's life, so doing produces the most utility. Once the soldier identifies what course of action promotes the most utility he must undertake it because to fail to do so would be to act wrongly. What is more, this is so for all utilitarians in similar situations. If throwing themselves onto a grenade promotes the greatest utility, then all soldiers in similar situations must undertake this course of action. In a similar situation, should the deonticist conclude that the right course of action in this situation is for the soldier to throw himself onto the grenade then this becomes their duty. To fail to do so would be to act wrongly, and this will also be the case for all deonticists in similar situations. For both utilitarianism and deonticism, once we acknowledge what would be the best course of moral action in a situation we act wrongly in not pursuing it.[7]

What justifies the partialist in criticising impartialism for not allowing us to favour our self-concerns over moral considerations is the impartialist's failure to appreciate how the content of our lives makes them worth living. Our activities and attachments satisfy our basic needs and interests, which are not simply those that are necessary for mere survival. We need a range of social relationships and have interests in aesthetic and creative activities, for example. What makes the brilliant scientist be of good character is the fact that they are morally good but also maintain and pursue a range of valuable non-moral activities. As well as satisfying their basic requirements, they make them more rounded, developed individuals, and it is this, along with their moral qualities, that means they are of good character and live well. It is not only the experiences of our activities and attachments that make life meaningful, but also our commitment to them. We pursue a number of activities which do not contribute to the fulfilment of our lives; it is our commitment to those that do which distinguishes them.

An impartialist can accept maintaining and pursuing non-moral activities and attachments, but they do not demonstrate any strong commitments to them. Whenever there is a conflict between our self-concerns and morally beneficial considerations, the impartialist requires the former to be forsaken. Instead, the impartialist demonstrates a commitment to being what Susan Wolf (1982) calls a 'moral saint'. Moral saints are not people of good character, as it is understood here, because they fail to appreciate the role and value of our self-concerns, and so do not understand what it is to pursue the good life.

Justice and elitism

While partialism maintains that self-concerns can take precedence over moral considerations, it must accept that there are limits to the activities and attachments that we can pursue. To fail to accept any restrictions on how we behave would be to reduce partialism to Max Stirner's anarchistic egoism (1995). The egoist does not undertake reciprocal friendships and altruistic actions because of the benefit they bring others, but only because they satisfy the egoist's desires.

What partialism needs, and what is fundamental to impartialism, is an account of justice.[8] It must explain why and when moral considerations do take precedence over our self-concerns. For example, that a person may leave their business to their children, thereby ensuring that it is a family-run business, is acceptable, but it is not acceptable for a person to give their cousin a post in a multinational corporation simply because of their familial relationship. The partialist must explain why this is so. It is not enough that the person who leaves their children their business is fulfilling a plan provided by one of their activities, or that they are benefiting one of their close attachments; these arguments are also open to the nepotist. In the case of the family-run business, the business is the property of the owner to leave to whomever they choose. The nepotist, on the other hand, acts on behalf of a multinational corporation and the post is not their property to give away as they please (Cottingham, 1997, pp. 9–11). The partialist needs an account of why such reasons as these are salient and justifiable.

The solution to the problem of justice for partialism is to be found by accepting statement (1). Yet, statement (1) places similar constraints on people as those provided by utilitarianism and deonticism. For statement (1) to restrain partialism and establish a different theory to impartialism it must claim that the restraints of utilitarianism and deonticism demand too much of people. If statement (1) rejects the Principle of Equality as a fundamental moral requirement it can make this claim. This modified statement (1) requires that the partialist act within the bounds of morality, which lacks the utilitarian and deonticist requirements of equality. It is the fundamental moral requirements that also protect partialism against the charge of promoting racism, nationalism and other forms of unfounded group concern, because the requirements establish when a concern is justified.

The moral requirements are also necessary to mitigate the charge of elitism against partialism to which statement (3) leaves it open. Statement (3) maintains that if someone does not pursue or maintain activities and attachments that reflect good character, their life will be less valuable than the life of someone who has a good character. For example, in contrast to the brilliant scientist, the lager lout does not demonstrate a valuable lifestyle, and so exhibits a deficient character because of their general dispositions, hence their life is less valuable than that of the brilliant scientist. It might be the case that to their loutish friends the lager lout is generous and kind, but to the rest of society they are disruptive. Their disruptive behaviour towards society undermines the dispositions on which the lager lout's generosity towards their friends is based. In accordance with statement (3), not only is the lager lout's life less valuable than the brilliant scientist's, but if the Normative Claim is retained, partialism implies that the lager lout may be treated differently, and possibly detrimentally, in comparison with someone whose life is more valuable, such as the brilliant scientist.

Were partialism concerned only with moral activities and attachments, the moral requirements might limit the charge of elitism. Certain moral dispositions are more favourable to the good life than others, as the case of the lager lout demonstrates. Nevertheless, partialism accepts that some non-moral activities are to be preferred to certain other non-moral activities because they reveal but also help to develop good character. To avoid the charge of elitism, the partialist must demonstrate that certain non-moral activities are better than others at promoting living well. The brilliant scientist has a good character, in contrast to the lager lout, because, I have assumed, they pursue a better range of non-moral activities. In some cases, it will be clear that the scientist's activities are better, for example, those affecting health, where drinking copious quantities of lager is unlikely to prove beneficial. In other cases, such as aesthetic appreciation, it is more difficult to demonstrate that some activities are more preferable, unless it can be shown empirically that they lead to a more rounded and developed character. In addition, the partialist does not seem to take into account the problem of moral luck (Cottingham, 1996, pp. 67, 72–3).[9] Many of the activities and attachments that people have are contingent on other factors, such as one's social environment and opportunities, but also possession of the relevant dispositions to pursue or maintain particular non-moral activities. This type of constitutive moral luck might deprive a person of the desire for certain activities and attachments, or the ability to pursue them.

If partialism accepts statement (1) (which no longer contains the Principle of Equality) it cannot provide the basis for morality. Partial concerns, as 'intensely personal commitments and preferential networks of mutual interdependence' (Cottingham, 1997, p. 7) might be thought to provide the necessary 'impulse' for our moral consideration of others beyond our immediate sphere of special attachments (p. 19). Without the fundamental moral requirements, however, such partial concerns could permit what is otherwise considered to be unjustifiable behaviour, such as that of the nepotist. What is more, the partialist cannot rule out that because of their character, the lager lout would be neglected and denied the protection of morality (p. 17). Our self-concerns and the 'networks of mutual interdependence' that are a feature of most people's lives will help us to cultivate and develop our moral sensitivities, but if we accept the role and importance of justice in morality, they cannot, on their own, form the basis of morality. What does form the basis of morality and why it is applicable to all human beings is the issue to which I now turn.

Human nature and equal respect

Partialism makes the important claim that people do not regard each others' lives as equally valuable. The life of a brilliant scientist, for example, is more valuable than the life of a lager lout. With statement (3) and the Normative Claim, partialism implies that the lager lout may be treated differently from the scientist. Furthermore, some accounts of partialism propose that partialist concerns can form the basis of morality, despite risking leaving some people outside of the moral sphere. This is not so with impartialist accounts of morality. Starting from a subjective perspective, where each individual considers their life as special, the impartialist claims that because all people find their lives special, people's lives are of equal value. When coupled with the Normative Claim, the Principle of Equality maintains that people should be valued and treated equally. Partialism and impartialism would seem incompatible, but the need for partialism to accept statement (1) points to a way in which it can support an understanding of the idea of human equality.

The Principle of Equal Respect

The sense of justice that partialism needs, and which impartialism implies, is a requirement of equal respect. Justice requires that no matter what differences exist between people's relationships, or the different values of people's lives, certain types of behaviour should not be undertaken. In order for impartialism and partialism to become compatible it is necessary to look outside of the content of a person's life, and towards facts that people have in common.

If equality is to get a grip anywhere it must be, as Williams (1962, p. 230–1) argues, at a basic level. One fact that unites all people is that they are human beings. Williams points out that the claim that human beings are equal in their humanity risks trivialising the conception of equality. Nevertheless, a consideration of the fact that what all human beings have in common is their humanity could serve to remind us of certain facts about human beings that are morally salient and which might have been forgotten (p. 232). All human beings have in common the fact that they have certain fundamental needs and interests. For example, we all need food and shelter, and are capable of experiencing pain. Such facts, Singer and Regan recognise, are common to many animals, human and nonhuman alike. But the basic requirements of human beings reflect those characteristics that are distinctive of people, such as our advanced rationality, complex relationships, and aesthetic and creative desires. It is the fundamental facts about human beings, our basic needs and interest, that provide the basis for the idea of human equality.

One fact that human beings have in common is their individual commitment to the activities and attachments that provide their lives with value and meaning. Many of these activities and attachments will satisfy our basic requirements, and there are plural ways in which people can satisfy them. Indeed, the ways in which our needs and interests are fulfilled provide people's lives with its distinctive pattern. While the content of people's lives will differ, they have in common the fact that they are committed, albeit in varying degrees, to the activities and attachments that shape and reflect their characters, and which make their lives meaningful and valuable.

It is in virtue of these fundamental facts about human beings that there is a requirement of equal respect. Respect towards others is demonstrated by behaving 'towards them as we morally ought to behave' (Raz, 2001, p. 126).[10] What constitutes morally appropriate behaviour is determined by the fundamental moral requirements, which are grounded upon the basic features of human beings. It is for this reason that we owe equal respect to all people, because, other things being equal, the basic features that ground morality are common to all human beings. It does not follow from this claim about equality that people should always be treated equally. How people should be treated will be determined by the fundamental moral requirements in particular circumstances and also by the fact that other things are not always equal, where some people do not possess all of the requisite characteristics about human beings, such as human embryos. It might be the case that people should be treated equally in a given situation, but we may have to give reasons, based on the morally salient properties of the situation, for so doing. In a similar way, we might be required to give reasons to justify treating people differently. For example, the owner of a family-run business may leave their business to their children, because they own it. The nepotist, however, should give their cousin the same opportunity as other candidates for the post because the nepotist acts on behalf of a multinational corporation. In both cases, a reason is provided to justify and explain the partial and impartial behaviour.[11]

The moral requirement that each person should be shown equal respect – that we should demonstrate appropriate moral behaviour towards every person – does not lead to the conclusion that every individual's life is of equal value. For example, Raz (2001, pp. 161–3) uses the evaluations of different paintings to explain the relationship between value and respect. It might be the case that one values the paintings of Raffaello Sanzio more than the works of Pablo Picasso. It does not follow from this, however, that one regards Picasso's works as being without intrinsic value. In a similar way, regarding the life of the brilliant scientist as being more valuable than that of the lager lout is still to recognise the value of the lager lout's life. We recognise that the lager lout's life is intrinsically valuable in virtue of the fact that, as a human being, they possess certain morally salient features, such as the ability to feel pain and their commitment to the content of their life. It is in virtue of these features that we owe them equal respect, even though we disapprove of their activities, which we may demonstrate in some suitable way. What constitutes morally appropriate behaviour towards any person will be determined by the context in which this behaviour takes place. The fact that the lager lout engages in disruptive behaviour will justify society's intervening in their activities. Nonetheless, the fact that the lager lout deserves equal respect means society's intervention must be a just one, where good reasons can be given for the response to their behaviour.

My account of equal respect can be summarised by restating statement (1):

> *The Principle of Equal Respect*: A person is justified in pursuing their activities and attachments except where to do so will fail to treat people respectfully.

In contrast to statement (1), the Principle of Equality and the Normative Claim are rejected as fundamental moral requirements. Statement (3) maintains that the value of a person's life is assessed according to their character. The fundamental moral requirements, however, are applicable to all people independently of their character, because they are grounded upon fundamental facts about human beings. How people are to behave towards one another will be influenced by, but is not reducible to, the values of the activities and attachments that they maintain and pursue.[12]

The Principle of Equal Respect provides an account of equality but it is more restricted than the Principle of Equality. Moreover, the Principle of Equal Respect does not resolve the conflict between partialism and impartialism. Rather, it maintains that the resolution of this conflict occurs at the level of the fundamental moral requirements. What is needed to resolve this conflict, when it arises, is an account of the fundamental requirements and how they determine what ought to be done.

Moral perception and particularity

What is missing from my account of partialism, and the role of the Principle of Equal Respect, is an account of moral epistemology. The partialist must explain why the actions of the nepotist are wrong and those of the owner of a family-run business are morally permissible. In addition, the Principle of Equal Respect does not explain what the fundamental moral requirements are or how they resolve issues of moral conflict. In the previous chapter I began to provide an account of moral epistemology, which will provide answers to these problems. It is through reason, that is, Aristotelian perception, that conflicts between rights and duties are resolvable, and it is through reason that moral agents come to discover the fundamental moral requirements. Furthermore, I maintain that Ross's account of intuitionism, which is the process of reasoning, together with his account of (retitled) *pro tanto* duties, provides an adequate account of how to resolve issues of moral conflict. It is only by bringing reason to bear upon the facts of a situation that we can decide whether partialism or impartialism is justified. Yet, Ross's theory of intuitionism and *pro tanto* duties requires further development and modification.

The account of moral epistemology that best explains how we recognise what is morally appropriate relies upon the three central features of facts, perception and the particulars of a situation.[13] A decision can only be made about the just nature of a situation when all the facts appropriate to making such a decision are known and considered. These facts must include features about human beings, that they have special attachments or that they can experience pain, for instance, and also features about the particular situation in question. Whether a fact provides a reason for an action or decision will depend upon the perception of a moral agent of the significance of that fact. Facts will have varying degrees of saliency according to the particular circumstances of a situation, where two situations are rarely (if ever) identical.

For example, that the lager lout in the burning house is the rescuer's sibling provides a reason for the rescuer to save the lager lout instead of the scientist. If the scientist were the only person who could save the world, this might provide a greater reason for the rescuer to save the scientist. Where the scientist can save the world, the reason for saving them instead of the sibling lager lout might appear obvious, and so the perception of the saliency of a reason for action is sometimes compared to the immediacy of realising a mathematical solution. This is not to imply that deciding to save the scientist would be an easy decision, and neither is it clear that to fail to do so would be wrong. Indeed, while it may be the case that some courses of moral action are obvious, perception often requires deliberation, and reservations about the decision, which is not synonymous with mathematical revelation.[14]

A difficulty for moral epistemology is the provision of an appropriate account of rules. If the account of moral epistemology followed here is concerned with specific situations, it is not clear what form moral rules must take. Universal principles fail to take account of the open-textured nature of particular situations, where in one situation φ-ing is permitted, but in another it is not (where φ and other Greek letters represent a salient moral property or verb of action).[15] The problem here is not simply specifying all the possible situations where φ-ing is or is not permitted, but of not knowing in which circumstances a rule might be called upon. As was noted, a specified right might be modified in the light of new cases. If it is possible to perceive when a rule needs a new specification, however, there is no need for a universally applicable rule, only a general rule and some epistemological account for determining how it is to be followed. This much is implied by Aristotle who, Roger Crisp (2000) argues, combines moral perceptivity with generalised rules:

> But let us take it as agreed in advance that every account of the actions we must do has to be stated in outline, not exactly. As we also said at the start, the type of accounts we demand should reflect the subject-matter; and questions about actions and expediency, like questions about health, have no fixed [and invariable answers].
>
> (Aristotle, 1988, 1104a1–5)

Indeed, this is in part Ross's solution (1930, p. 32) to the problem, but it describes moral action from the opposite direction.[16] On Ross's account, a property of a situation determines what should be done. An agent knows the moral saliency of this property because they perceive it, and this perception also makes the agent aware that the property will be relevant in all similar situations. This account of the property gives rise to Ross's generalised (retitled) *pro tanto* moral principles, even though the process of perceiving which properties are morally salient is the same in every moral situation. Moral principles are always known through intuitive induction of particular moral properties to generalised moral principles, rather than deductively from a general principle to the moral property.

Ross requires that if a property makes a moral difference in one case it will always make a difference and in the same way in any other situation in which it occurs, unless there is another property which makes more of a difference. If pleasure provides a reason in favour of performing some action, for example, it always provides a reason in favour of performing some action, even when another property is more salient in a particular circumstance. Whether pleasure is a salient feature in particular situations depends upon the perception of the agent. For example, if A must choose whether to aid B or keep their promise to meet C, the pleasure of meeting C might count in favour of keeping their promise to meet C, but A might perceive the saliency of aiding B as more stringent. There are examples, however, where pleasure is a reason for not doing something, such as the pleasure gained by fox hunters. If the cruelty of fox hunting is mitigated by its alleged necessity, this is undermined by any pleasure gained by the hunters from the killing of foxes. Indeed, any pleasure gained from performing cruel acts tends to suggest even greater reasons for not doing them (Dancy, 1993b, p. 61).

A further issue is the way in which different properties of dissimilar situations will affect the saliency of reasons, in what Jonathan Dancy (1993b) refers to as the holism of reasons.[17] If there are two properties present, φ and ψ, in a particular situation, φ might be more salient leading to φ-ing, with ψ a defeated reason. In a theory of general duties, if φ-ing is the right course of action, ψ plays no function in the action. Yet, in a different situation, with properties φ and π, both might be salient, and while φ might be more salient, leading to φ-ing, property π does not disappear. Rather, π alters the saliency of φ and affects the act of φ-ing, occasioning the feeling of regret that π is not acted upon, for instance.[18] An example of the effect that different moral properties can have upon the saliency of reasons for action is Jim's decision to kill one of the Indians and the regret he would feel in so doing (were he to decide that this is right course of action). Holism about reasons is an attempt to explain how different situations affect moral properties in different ways, but also allow different moral properties to operate in unison, thereby affecting moral motivation.

The problem with generalised moral rules is the need for a property to act logically in the same way in each situation in which it occurs. If it did not act the same way, a general account of it could not be constructed. Ross's account of (retitled) *pro tanto* duties provides an epistemological account of how a pluralist account of general moral principles can function and avoid contradictions. Dancy (1983, p. 543), however, claims that an important problem for Ross's account of moral conflict is Ross's metaphysical need for rules. Despite his epistemological account of moral conflict, Ross insists that moral decisions are supported by generalised moral rules. One important justification for generalised moral rules is the usefulness of bringing past experiences to bear on solving present conflicts. Ross's account of moral epistemology might appear to exclude the use of past experiences because of his claim that the saliency of a feature is perceived in each situation. Nevertheless, moral perceptivity is guided by past experiences, particularly in shaping the character of an agent. Dancy draws a comparison between moral perceptivity and Wittgensteinian rule following (p. 545). Wittgenstein (1967, sects 185–242) describes rule following, not as following a set of instructions, which binds the course of an action to previous actions, but as grasping the nature of a rule. Rule following is a skill that is learnt through experience, not simply by following instructions. It is grasping a rule that allows one to set a precedent, and setting a precedent might require that a feature, such as pleasure, acts in an opposing way to previous examples. Wittgensteinian rule following does not refer to generalised rules, but to the saliency of properties in specific situations. Contrary to Crisp (2000, pp. 27–32), who maintains that Aristotle adopts a generalist approach to moral rules ('every account of the actions we must do has to be stated in outline') the notion of perception and what it implies suggests Aristotle has a Wittgensteinian approach to rules.[19] This becomes apparent when we consider the role of character in moral deliberation.

The role of character

An important criticism of Aristotelian perceptivity is its assumption that people will perceive the same situation in the same way. To ground morality on human characteristics provides some source of continuity between people but, despite this, there remains considerable room for dispute. More problematic for this account of moral epistemology is the issue of genuine evil, where people are aware of certain facts but deliberately act badly. If moral perception is the ability to reason, all moral agents ought to perceive right conduct and avoid evil, but this is not the case.[20]

H. A. Prichard's response (1912, pp. 9–10, n. 9) to these criticisms is to suggest that where differences of opinion arise, they do so because one or all parties involved are not sufficiently developed moral beings; or they might have failed to consider adequately the consequences of any proposed action; or the disputing parties might be thoughtless, and have failed to consider all the facts and how they relate; or a combination of all three of these reasons. What constitutes a developed moral being, one who is capable of considering the consequences of a proposed course of action and recognising the salient facts of a situation, is someone who is of a good, that is, virtuous, moral character. As John McDowell argues, '[o]ccasion by occasion, one knows what to do, if one does, not by applying universal principles but by being a certain kind of person: one who sees situations in a certain distinctive way' (1979, p. 347). In order for a person to perceive that an action is morally justified, permissible, wrong or supererogatory, they must be predisposed to the morally salient properties of a given situation, the perception of which leads to the appropriate conduct. Thus, for a person to perceive that the salient property of a situation is kindness, and therefore that they should be kind, this person must be predisposed to acts of kindness.[21]

Statement (3) promotes a similar view to McDowell's when it states that a person's character reflects and is shaped by the activities and attachments that they maintain and pursue (which give their life value). A person of good character can live the good life because their activities and attachments contribute to living well. For this to be so, an individual must maintain and pursue a range of appropriate moral and non-moral activities and attachments. A moral saint does not live well because they focus only on moral activities. In his account of an 'intelligent' person, understood here as a person of good character, Aristotle declares, '[i]t seems proper, then, to an intelligent person to be able to deliberate finely about what is good and beneficial for himself, not about some restricted area – e.g. about what promotes health or strength – but about what promotes living well in general' (1985, 1140a25–8).

The Principle of Equal Respect requires that the notion of 'living well in general' should have an additional meaning, namely that the good life is not obtained in isolation from others and the society in which one lives. We are, as Heidegger describes it, Being-with-others; we are members of a community, one that consist of many different levels, from our families to the worldwide community. This is a fact that shapes us and our needs and interests, and as such provides a basic fact that grounds morality and the pursuit of the good life. Many, if not most of our activities and attachments will affect one or more people, and some will affect the wider community. For a person of good character to deliberate properly, they must consider all the available facts of the situation in question and the wider implications of their actions.

Partialism cannot provide the grounds for morality because it risks permitting an unjust society. Some people may be left outside of the network of partial concerns, and their treatment, which may be unfair, will not necessarily be of concern to those within the network. On the account of the good life supported here, such a society does not suggest that the conditions necessary for living well exist.[22] To disregard the unfairness of society, when one's life is seemingly good, means those people who apparently live well lack certain moral dispositions to charity and justice, for example, and so are not living well. This should not be taken to imply that any unfairness in society prevents everyone from living well. What prevents the pursuit of the good life is to ignore this unfairness and not attempt, in some way, to address it.

The fundamental moral requirements

The moral virtues are analogous to what I refer to as the fundamental moral requirements. While these requirements are grounded upon facts about human beings, our environmental situation and human needs and interests, it is important to emphasise that the moral requirements are not reducible to these facts. Neither are the moral requirements parts of a framework or system ensuring the fair and just maintenance and pursuit of our activities and attachments. Just as it is part of the human condition to have certain basic requirements, so it is a distinctive feature of this condition that human beings, other things being equal, have the capacity to reason. Thus a person of good character is capable of perceiving moral properties from the facts of a particular situation and of acting upon the saliency of them.

What unites the virtues and duties is the fact that they represent appropriate responses to the moral properties of a situation. A salient property might require that one be kind, charitable or keep a promise, whichever virtue or obligation is fitting. Virtues and duties must be understood as involving a Wittgensteinian sense of rule following. To follow the moral duty of non-maleficence or the moral virtue of courage is to grasp what the concepts of harm and courage entail and how to apply them in circumstances where they are relevant. To be of good character is to grasp how to recognise and respond to fundamental moral concerns in new situations. This is a knowing-how not a knowing-that knowledge and can only be obtained through experience, not simply through learning by instruction. In order for people to develop good characters, they must live in a culture that promotes the moral and non-moral virtues that contribute to living well. The cultural background provides the conditions and values for developing the necessary predispositions of good character. If moral rules have a place, it is as a form of communicating between moral agents and as tools for instruction and teaching, and importantly, for framing laws.

There are many who are sceptical of uniting deontic accounts of duties and obligations with the moral virtues.[23] Given my account of moral epistemology, the moral virtues and moral obligations are not antithetical ideas but complementary. The moral virtues define certain qualities that a person of good character must possess if they are to perceive right conduct. Rights, duties and obligations do not define or provide a proper sense of these qualities. G. E. M. Anscombe (1958) and Williams (1993) also observe that the terminology of duties and obligations fails to encompass the inexactness of certain relationships, or the fact that an action might be blameworthy but not wrong, or praiseworthy but not right. While the moral virtues have these advantages they cannot encompass the specific nature of moral duties and obligations, nor can they encompass all of the fundamental moral relationships of duties and obligations.[24] There is a sense in which certain attachments do place those involved under obligations. A parent has a moral obligation to care for their children, for example, but describing this moral responsibility in terms of the virtues does not capture its nature as a requirement. If ethical discourse is to capture and appreciate the saliency of moral properties, it must be open both to the virtues and deontic notions of duties, obligations and rights.

Equality and longevity

In Chapter 4, I outlined three alternatives to the threat of overpopulation that increases in life spans might cause. The traditional solution to the prospects of overpopulation is to reduce birth rates, but this may not be sufficient to accommodate increases in life spans. The remaining alternatives are either to restrict increases in life spans or to abandon maintaining an optimal population level. The Principle of Equal Respect, along with my discussion of moral epistemology, clarifies the issues these two options raise, but does not indicate which one should be taken.

Maintaining equal access to resources

My purpose in examining the idea of human equality is the presumption that every individual should have the equal opportunity to flourish. The Principle of Equal Respect explains in more detail why this is so. Other things being equal, we all have the same basic requirements, and how we meet them shapes and provides our lives with meaning and value. If we are each to have the opportunity to pursue the good life, we must have equal access to the resources that we need to satisfy our basic needs and interests. There are a number of ways in which we can flourish, each of which may require different quantities and qualities of resources. Furthermore, while we are equal in having similar needs and interests, their nature will not be the same. The healthcare needs of some for maintaining normal functioning will be greater than for others, for example.

Increases in life spans threaten to cause overpopulation in two ways. First, they will involve an increase in the number of people. Second, longer life spans will increase the demand for resources made by each individual. Increases in longevity will be achievable only with developments in medicine, which means that each person will require more healthcare resources. It is not obvious that only the elderly will require more healthcare resources. Increases in longevity will involve repairing the damage that causes the ageing process, but it may also involve preventing the cellular damage that causes ageing in the young. Nonetheless, the elderly may require more resources if, by improving the quality of their health, they are able to purse a wider range of activities than is presently the case.

With increases in longevity leading to more people and a greater demand by each person for resources, attempting to maintain the equal opportunity to flourish may force society into a state of affairs that does not sit on the Five Dimensions Theory's indifference curve. Such a state of affairs would involve real overpopulation, and so no individual would have the opportunity to flourish. What is more, real overpopulation may restrict increases in life spans because people would not have access to the resources they need in order to maintain and restore the quality of their health.

Whether such an outcome would occur depends both on how significant are the increases in life spans and how resources are distributed within society. So long as progress is made in treating and preventing the ageing process and the diseases and disorders associated with it, other things being equal, life spans will continue to increase. Real overpopulation may occur because movement is accepted along the axis of the Five Dimensions Theory's indifference curve concerning the number of people in society without making sufficient trade-offs with other values, specifically the quality of life.

To accommodate increases in life spans while retaining equal access to the resources necessary for having the opportunity to flourish, it may be necessary for society to adopt a different position on the Five Dimensions Theory's indifference curve. A new position will involve a reduction in the range and quality of the types of activities people can pursue, thereby releasing resources for the healthcare that is necessary for sustaining longer life spans. For example, we might accept a mediocre performance of Mozart rather than an exemplary performance if this enables us to maintain the quality of our health in later life, with the consequence of increasing longevity. While we might accept changes to the range of activities we can pursue and a decline in the quality of our experiences, our activities and attachments must be sufficiently fulfilling in order to make life valuable and open to the pursuit of the good life. A mediocre performance of Mozart may be acceptable, but a poor performance will not.

As was noted earlier, longevity is not valuable in isolation from the other goods of life but is a structural feature of the good life. What makes life good are the experiences we obtain from our activities and attachments, and so long as they are good, all things considered, we want them to continue. If it is necessary to curtail the quality of the content of our lives to such an extent that life is no longer good, longer life spans would be undesirable. Furthermore, a situation where we were no longer in a position to pursue the activities and attachments that make life good would lack the opportunity to flourish. Society would have moved to a state of affairs that does not sit on the Five Dimensions Theory's indifference curve because the trade-off requiring sacrifices in the quality of life would be too great.

Restricting the length of life

The detrimental consequences for the pursuit of the good life that retaining equal access to resources creates, suggests a second solution, that of limiting how long people can live. The motivation for this solution is that it is better and fairer to have the opportunity to flourish for a certain length of life than never to have the opportunity to flourish (assuming that retaining an equal access to resources will prevent people from having an opportunity to flourish).

The decision to limit the length of life does not imply that life spans cannot increase. Once the decision has been taken to limit longevity, society is in a position to choose which place on the indifference curve to adopt. A state of affairs might be chosen that does not allow for increases in average life expectancy or only allows for modest increases in order to ensure a wide range of opportunities for flourishing and a higher quality of experiences. The alternative is to allow for greater increases in life expectancy but accept a smaller range of opportunities for flourishing with lower-quality experiences. Unlike the previous option of retaining equal access to resources, where the lack of the opportunity to flourish may in fact restrict the length of life, the decision to limit longevity is undertaken to ensure, insofar as it is possible, that every individual has the opportunity to flourish for a specific duration of life.

A restriction on the length of life challenges the pursuit of the good life in at least three ways. First, it will require a different understanding of genuine healthcare needs. The imposition of a normal life span will involve denying those people who live longer than a specific age restriction access to the broad range of healthcare. The healthcare needs of such people, defined in terms of normal functioning, will not change; those people who live longer than the normal life span will still have an interest in maintaining the quality of their health. As a consequence, the role and purpose of medicine will need to change. Rather than maintaining species-typical normal functioning, understood as the practical functions of cells, tissues and organs, it may be necessary to adopt a teleological account of normal functioning. What range of healthcare this will involve throughout the course of life is unclear, and will depend largely upon how long is the imposed normal life span. The changing role of medicine would not necessarily remove the entitlement to healthcare, although it would involve altering the range of healthcare options to which a person is entitled at different stages of their life. For instance, any person who lives longer than the normal life span may still be entitled to palliative care, while there might be restrictions on access to treatments for Alzheimer's disease and cancer, for example, prior to the limit of the normal life span.

Second, a restriction on the length of life alters the appreciation of the misfortune of death from the objective point of view and challenges the value of longevity. The establishment of a normal life span implies that although it is good that a good life continues, it is good only if it continues for a certain time. As was noted, the tedium of immortality also recognises that it is possible to live too long, but this is for the subjective reason that that the good life ceases to be fulfilling. In contrast, limiting life spans is the result of the objective, impartial claim that it is better for society, presently and in the future, that the good life is curtailed.

Although Tolstoy lived for as long as he might have reasonably expected and his death did not (presumably) deprive him of many future goods, it can still have been a misfortune for him. The idea of a limit to longevity, be it natural or otherwise, is not a feature of the subjective point of view. So long as Tolstoy found his life fulfilling he would have wanted to continue living because this is an essential feature of the good life. It is on the basis of the subjective point of view that we can recognise from the objective point of view that Tolstoy's death is a misfortune for him. A restriction on the length of a good life, however, creates an ambiguity to the objective perspective of the misfortune of death. On the one hand, we can recognise that from the subjective point of view, death will be a misfortune for those who live for as long as the age restriction, if their life is good. On the other hand, while it is good that a good life continues, it is good only if it continues for a limited time. In creating this ambiguity, curtailing the good life challenges the value of longevity, which is a fundamental feature of living well.

Third, establishing a normal life span will challenge the idea that human beings are fundamentally equal. It was noted earlier that Rawls maintains that inequality may be justifiable if the whole community benefits from it and it may be the case that limiting longevity is the best option available. Nonetheless, it is important to be clear about what this implies for the idea of human equality, and also to question whether such inequality would be fair. The ideal of equality maintains that no one individual is more valuable than another. The Principle of Equal Respect explains why, at a basic level, we are equal and hence why, for example, the lager lout, despite having a poor character, is no less valuable than the brilliant scientist. We each have in common the need to pursue the good life, which grounds the claim that every individual should have an equal opportunity to flourish.

The reason for restricting increases in life spans is that it is thought to be fairer to ensure that every individual has the opportunity to flourish for a certain number of years rather than risk no one having this opportunity. The fair innings argument attempts to justify age rationing by appealing to the natural life span, a biological fact about human beings that can explain why the basic requirements of those who live longer than a fair innings are less valuable than those who have yet to achieve it. Without a natural life span, an alternative argument is needed to explain why rationing healthcare according to age, and so restricting life spans, is fair.

The difficulty for age rationing is that it is not obvious that it is fair. The equal opportunity to flourish is grounded upon the basic needs and interests of human beings, and these will be no different for people who live longer than a socially imposed restriction on longevity than those who have yet to reach it. As a consequence, it wrongs no one to respect the opportunity to flourish of anyone who lives longer than the age restriction, except if we abandon the Principle of Equal Respect. To do so, involves challenging a fundamental feature of the good life. What is more, an assumption has been made that the Principle of Equal Respect would be replaced by statement (2), but it is not obvious that this would be the case. Once the normative connection between our basic requirements and the opportunity to flourish has been broken, there is no reason why an alternative basis for the distribution of resources cannot be established, for example, based purely on wealth.

7
Longevity and the Good Life

A number of different issues have been covered, and it will be worthwhile pulling together the central strands of my argument. The starting point of this discussion was to consider whether increasing human life spans constitutes a form of medical enhancement, which is appropriate and useful for a number of reasons. First, human beings along with most other animals have a biologically limited life span, and gerontological theories explain that this is caused by the ageing process. The use of developments in medicine to prevent or retard the ageing process to increase life spans beyond this limit, therefore, would appear to constitute a medical enhancement. Second, there is considerable debate about why and what types of enhancements should be prevented. This debate provides a useful framework for analysing some of the ethical issues that increasing life spans will raise. Part of my aim has been to place the discussion of increasing life spans within the broader debate about technological developments and the relationship between technology and our values. Medical enhancements are, by their very nature, a result of technological innovations, which makes the framework for analysing their ethical implications particularly relevant.

An initial conclusion to draw from my argument is that increasing life spans does not constitute a medical enhancement. As I understand the idea, medical enhancements identify interventions that improve the human form or functions in ways that are unacceptable. The concept of a medical enhancement is not, however, straightforwardly about augmenting human beings; some descriptive enhancements are desirable, such as immunisations. Rather, medical enhancements are those interventions that, by altering the human form or functions, do not satisfy genuine medical needs. An extension of human longevity will not constitute a descriptive enhancement of people because, I argue, there is no natural limit to longevity. This is not to dispute that the ageing process provides a biological constraint on the life span. How long we can live, however, is dependent upon human activities and the extent to which they can and do overcome such biological limitations. The ageing process, the gradual accumulation of damage at the molecular and cellular level, is but one of many impediments to continuing to live in good health, be they social, economic, environmental or biological. So long as we value good health, and we view maintaining good health as a genuine medical need, ageing will be regarded as a disease because it involves the abnormal practical functioning of our bodies' systems and processes. What is more, to be successful at preventing or treating ageing and its associated diseases and disorders will lead to increases in life spans.

A brief assessment of the value of a good quality of health indicates the value of longevity. Good health, understood as normal species-typical functioning, is valuable because it enables us to fulfil and maintain our activities and attachments. Yet, increases in life spans are more than simply an incidental but not undesirable consequence of endeavours to maintain and restore the quality of health. The assessment of the misfortune of death emphasises that when life is good, other things being equal, it is better to have a longer life than a shorter one. When we are in reasonably good health and in circumstances where we are able to experience and find rewarding the fulfilment of our categorical desires, we want to continue living. As a consequence, the pursuit of longer life spans is a valuable endeavour because longevity is a structural feature of the good life. It is structural rather than contributory because we do not want longevity itself, but want those features that are constitutive of the good life, such as maintaining friendships or aesthetic activities, to continue when we find them rewarding.

How an outcome is achieved can contribute to its value; but just as this contribution can be positive, so it can also be negative. The way life spans are increased may detract from the value of making people live longer, particularly if the means for so doing involves wrongdoing. A wide range of different biomedicines will be involved in increasing longevity, but my concern throughout has been the way in which longer life spans will affect our moral relationships. The pursuit of the good life is not undertaken in isolation. It is a fact about human beings that we are Being-with-others, and along with other facts about us, it grounds morality and living well. The use of nonhuman animals in the development of medicine undermines the integrity of our moral convictions about what constitutes morally appropriate behaviour, specifically in relation to the infliction of suffering. Our moral convictions demonstrate possession of the moral virtues and are expressed through our behaviour. By undertaking, supporting or tolerating activities, such as animal experimentation, we undermine the integrity of our commitment to certain types of moral behaviour that contribute to living well. The fact that we are Being-with-others means there is an important social element to living well. To live in a society that tolerates and benefits from animal experimentation challenges the pursuit of the good life for every member of society.

Why animal experimentation is undertaken can be explained by the technological understanding of Being. To understand the world, including animals and human beings, technologically is to view it as something that needs controlling, but also as a resource and a tool for achieving this order. The process of ordering serves to prevent the fragility of life that arises because of the contingent aspects of our existence. This is not to propose that we should avoid seeking to prevent some of the vulnerability of our lives, but that when we do, it must be in harmony with the pursuit of the good life. Activities that challenge the integrity of our commitment to certain types of appropriate moral behaviour are not in harmony with living well.

Present medical practices challenge the pursuit of the good life, but this does not mean longer life spans cannot be achieved without involving wrongdoing. There are alternatives to animal experimentation, which could prove to be sufficient for producing the necessary developments in biotechnology to add quality years to people's lives. In contrast, it may be impossible to avoid the consequences of increasing life spans for the fair distribution of resources. So long as we value good health and are reasonably successful at maintaining and restoring it, increases in life spans will be inevitable.

One response to the pressure extending life spans will place on the fair distribution of resources is to continue allowing people equal access to the resources they need in order to have the opportunity to flourish. With scarce resources and an increase in demand for them, it may be the case that people receive insufficient resources to meet their needs and so they will lack the opportunity to pursue the good life. The alternative response is to prevent longer life spans, or allow a limited increase in longevity, by establishing a normal life span. Such a policy may also be detrimental to flourishing by threatening values that are fundamental to living well. First, limiting longevity means not meeting the genuine medical needs of those people who live longer than the normal life span, which requires reassessing the role and purpose of medicine and the value of good health. Second, establishing a normal life span challenges the idea that longevity is an important structural value of the good life, and in so doing gives rise to an ambiguity about the misfortune of death. Third, placing a limit on how long we can live disputes the idea that human beings are fundamentally equal.

The consequences of increasing life spans are that they will require unacceptable trade-offs between our values. Rather than moving society to a different state of affairs on the indifference curve of the Five Dimensions Theory, greater longevity will move society off the indifference curve. The accommodation of longer life spans will require repositioning the indifference curve by reassessing what it is for human beings to live a good life. Whether this would be possible is doubtful. The values in question relate to good health, longevity, equality and the quality of life, which are grounded on our basic needs and interests. Moreover, throughout an assumption has been made that it is possible for a society to move to a different position on the indifference curve and adopt a new ideal of the good life. In principle this is possible, but in practice it will prove difficult. Each state of affairs on the indifference curve represents a particular understanding of the good life, one that will be embedded in and shaped by a specific cultural background. The clearing provides our values, and although it is possible to alter it from within, so doing will be a gradual process taking a considerable time. What is more, given the specific historical, social, political and economic features of the background it is difficult to conceive of a radical alteration taking place.

The somewhat pessimistic conclusions of my argument rely upon a number of assumptions about the lack of available options for preventing overpopulation and the scarcity of resources. I argue that a minimum number of children born in a given period is desirable, not least because of human nature, the value of children and their attendant culture, but also for economic reasons. I make no claim about what this number might be, and, as such, there could remain considerable scope for the reduction of the birth rate to accommodate increases in life spans. Natural resources are scarce because they are finite, but this does not show that they are sufficiently scarce to bring about the difficulties I describe. A state of affairs might exist whereby society has yet to reach its optimal population level and the population can grow without requiring a reduction of the available resources for each individual. Alterations to economic institutions and systems might also make more resources, and in particular healthcare, available. Furthermore, I provided a sceptical appraisal of a technological solution to overpopulation because such solutions to social and economic difficulties often create new problems. It is not obvious that this is an inherent feature of technology and so it need not produce new difficulties when solving older ones. Advances in the development of new and future medicines might be such that the costs of medicine are greatly reduced, which by improving the quality of our health in later life could remove some of the costs and resources involved in caring for the elderly.

The extent to which increases in life spans could threaten the pursuit of the good life will depend on our present social, economic, cultural and environmental conditions, and not least how great such increases will be. It is unlikely that there will be increases in life expectancy as dramatic as those experienced over the course of the previous century. Present and future increases are reliant upon advances in biomedicine that can repair and prevent the ageing process and the availability of resources with which to develop and bring these technologies to fruition. Nonetheless, unless it becomes impossible to find technological solutions to ageing, or to fund them, life spans are likely to increase, albeit gradually. While slow but steady increases in longevity might not pose an immediate threat to well being, without some means of finding, making available or using our resources more efficiently, and also of finding a way of accommodating a growing population, the problems of distributing resources fairly are likely to occur.

What my argument emphasises is that there is a fundamental tension within and between our values, which implies that it is possible to promote them to excess and to the detriment of the good life. The tedium of immortality provides one example of this. While it is better for our lives to continue when they are good, it is only good for them to continue for a limited duration: it is possible to live too long. To live too long will inure us to the fragility of life, the contingencies that shape, directly or indirectly, our values. The tension in and between our values is made more evident by the difficulties that longer life spans will cause for the fair distribution of resources. Greater longevity will be a result of supporting the value of good health and the quality of life, but if we are successful at promoting these values, they will become conflicted. As the argument about curtailing life spans demonstrates, because they are our values we cannot easily, if at all, seek to stop from promoting them.

The question then arises about how and why a situation can arise whereby we promote our values to excess, and attention turns to the means by which we sponsor them. The role of technology in allowing us to realise many of our values, and the detrimental consequences that can arise or it is feared might arise as a result, has led some, such as

Jonas and Heidegger, to question the relationship that exists between technology and the pursuit of the good life. Jonas's argument that it is an intrinsic feature of technology to distort and determine our objectives is difficult to sustain. Technology is essentially neutral; it is the way in which human beings use technologies that has consequences for living well. This is one of the central points of Heidegger's analysis. His concern is not with technologies as such but a specific understanding of the world that views it as a resource and a tool for realising our values.

Increases in life spans are and will be the result of practices that seek to control some of the contingent features of our lives, specifically to provide good health and prolong our lives so that we can maintain and pursue the activities and attachments that make our lives good. As such, they do not appear to involve a technological understanding of Being, because the promotion of these values is in harmony with the pursuit of the good life. The detrimental consequences for well being that could arise from longer life spans suggest this is not the case. A criticism of Heidegger's account of the technological understanding of Being is that, while he proposes that it represents a dominant sub-culture of the modern age, it in fact continues a long debate about how much control we should have over the contingencies of life. What is not in dispute is that human beings do intervene in the world, including our biology, in order to pursue the good life. The argument denying that there is a natural life span points to the broad way in which this is the case with, for example, improvements in sanitation, agriculture and medicine all contributing to enabling us to live longer, healthier lives. Increases in life spans continue this process and do not represent a technological understanding of Being, at least as I have interpreted it in relation to living well. Rather, the tension that exists within and between our values that allows them to be promoted to excess points to another way in which living well is fragile. It is a vulnerability that our success at living longer, healthier lives will emphasise and intensify, and in so doing, could ultimately undermine the pursuit of the good life.

Notes

Introduction

1. Gerald Gruman (1966) provides a detailed description of various mythical, folk and historical searches for a way to live longer; Carol Haber (2004) discusses more recent endeavours, contrasting them with those of the Renaissance.
2. A position statement by S. Jay Olshansky et al. (2002) warns against popular claims that it will soon be possible to achieve considerable increases in longevity.

1 Longevity, Technology and Humanistic Values

1. The biblical claim that people will live for 70 years can be found in Psalms 90:10. This verse also maintains that some people, as result of their physical robustness, can expect to live for 80 years. This contrasts with Genesis 6:3, which maintains that human beings will live for 120 years.
2. Tom Kirkwood (2000, pp. 5–7) draws on data from the Registrar-General's report for England and Wales to compare life expectancy between the 1880s and 1990s, and show that life expectancy has risen from 46 years to 76 years. Olshansky et al. (2002, pp. B292–3) compare the average life expectancy of 47 years in the developed world in the year 1900 with the 77-year life expectancy of children born in the USA in 2002.
3. Bruce Carnes et al. (2003) do not draw the analogy with a machine, but, as I understand their idea of a biological warranty period, it lends itself to this. The analogy becomes more pertinent later in the chapter when considering the evolutionary basis for ageing.
4. I draw on the definition provided by John Maynard Smith (1962, p. 115) and Kirkwood's use (2000, p. 35) of a slightly rephrased definition by Maynard Smith.
5. Kirkwood (2000, p. 33) identifies the rate of ageing with the actuary, Benjamin Gompertz, who in 1825 identified the eight-year doubling of the probability of mortality. Steven Austad (1997, p. 11) identifies the rate of ageing with the more recent work of the neurobiologist Caleb Finch. Austad also observes that the rate of ageing is between seven and ten years, depending on various environmental factors. The rate of ageing has varied throughout history from between seven years and 26 years (Austad, 1997, p. 38). Not all animal species age. Kirkwood (2000, pp. 36–7) observes that sea anemones and freshwater hydra appear to be immortal.
6. Kirkwood (2005, p. 438) describes Peter Medawar's theory as the Mutation Accumulation Theory. Kirkwood (2000, pp. 78–9) provides a useful summary of this idea.
7. The elucidation of Kirkwood's Disposable Soma Theory draws on a number of his works (1977, 1997, 2000, 2005, 2008; Kirkwood and Austad, 2000;

Kirkwood and Rose, 1991). His ideas were used to develop the notion of a biological warranty period (Carnes et al., 2003, p. 32).

8. For the influence of August Weismann on Kirkwood's ideas, see Kirkwood (2000); and for Weismann's influence on gerontology, see Kirkwood and Thomas Cremer (1982). Kirkwood and Cremer (1982, p. 112) and Kirkwood (2000, p. 55) describe the germ-line as immortal.

9. Kirkwood (2000, 2005) provides details of the type of damage that is experienced at the cellular level. Austad (1997) provides a general description of the possible mechanisms of the ageing process, while Evan Hadley et al. (2005) outline the range of research areas into ageing therapies.

10. Haber (2004) claims that throughout the nineteenth century scientific discoveries defined old age as a disease.

11. Christopher Boorse (1975, p. 57) recognises that what is natural is also open to evaluation, but maintains that a purely descriptive account of the natural design of a function can be given.

12. Richard Norman (1996) and Russell Blackford (2006) describe various problems that the argument from nature faces.

13. Haber (2004, p. 516), when assessing literature on longevity from the eighteenth century, refers to Luigi Cornaro's claim that there are many virtues to ageing, if only it could be accompanied by good health.

14. Eric Jeungst et al. (2003) also consider whether increases in longevity are a medical enhancement.

15. The wide range of issues is discussed in Erik Parens (1998).

16. Boorse (1977) identifies normal functions as the typical statistical range for a reference class. This class need not be a species but, to take into account changes in functions over time, specific age groups and sexes.

17. On enhancements and how our vulnerability helps to shape our values, see Gerald McKenny (1998). Martha Nussbaum (1986) provides a fuller account of the importance of the fragility of goodness.

18. Thomas Nagel (1971) considers various responses to the sense of the absurd.

19. Sigmund Freud (1920, p. 317) describes the propulsion towards death as arising out of Necessity, that is, a law of nature.

20. Page numbers for Martin Heidegger (1962) refer to those of the translation. I draw on Stephen Mulhall (1996, pp. 114–20) for this account of an authentic approach towards death.

21. I utilise Joseph Raz's description (2001) of our activities and attachments as the content of our lives.

22. Galen Strawson (2004) provides an insightful critique of the idea that we should seek consciously to provide our lives with a distinctive narrative.

23. A distinction is drawn between being and Being. Human beings interact with a number of entities in different situations where each interaction and situation reveals an aspect of the entity's nature. In so doing, our interaction reveals what it is for an entity to be and for us to comprehend what they are. Being (upper case) refers to this understanding, that is, to the Being of being. For Heidegger, this ability to comprehend the nature of entities is an implicit capacity of human beings (Mulhall, 1996, p. 3).

24. Heidegger's account of Dasein can be found in Heidegger (1962) and the clearing in Heidegger (1971). For this elucidation of his account of Dasein and 'throwness', I have also drawn on Hubert Dreyfus (1991), Michael

Inwood (1997) and Mulhall (1996). For the account of the clearing, I am indebted to Dreyfus (1993) and George Steiner (1989).

25. Michael Oakeshott (1948), for example, is critical of rationalism, specifically in relation to politics, which presumes that the only valid knowledge is that which can be taught by instruction. Rationalism is the search for a universal set of rules that can be applied with certainty. In contrast, Oakeshott argues that practice, participating in the activities of a culture, imparts the kind of knowledge that cannot be taught, such as with Dreyfus's example of how far to stand from one another. Oakeshott's criticism of rationalism has echoes of Heidegger's critique of technology.

26. Peter Winch (1964) and Brenda Almond (1999) identify the commonality of such facts across human societies as grounds for some degree of commensurability between different cultures.

27. John Mackie (1977) provides an account of morality along these lines.

28. Heidegger (1962, pp. 91–148) distinguishes between objects that are ready-to-hand, such as a tools and those that are present-to-hand, objects that we contemplate. Don Ihde (1979) discuses this distinction.

29. John Cottingham (1998) assesses the role of reason and rationalism in the pursuit of the good life and considers Heidegger's objections in the broader context of this debate.

30. Dreyfus (1993) points to similarities between Heidegger's account of the clearing and Thomas Kuhn's description (1996) of scientific paradigms and the way that scientific revolutions leading to paradigm shifts occur when science seeks to accommodate the anomalies of a particular paradigm.

31. Norman (1996) provides an account of the argument from nature, which is similar in many respects to Heidegger's description of the clearing and the role it plays with respect to our values. Blackford (2006) describes Norman's argument from nature as at best one supported by a particular group in society (just as the technological understanding of Being represents a sub-cultural view).

2 The Misfortune of Death

1. Christopher Belshaw (2009) discusses in detail the various definitions of death. Eric Olson (1997) and David DeGrazia (2005) support an account of personal identity based on the living body; Jeff McMahan (2002) supports an account for which the criterion of personal identity is psychological continuity. McMahan (2002, p. 424) argues that what matters to us is our psychologies and, although our existence ends when our consciousness irreversibly ceases, our organism may remain and may continue to be alive. This dualism will require two definitions of death.

2. For my description of the role of the body for personal identity, I have been influenced by Bernard Williams (1956), who draws on Ludwig Wittgenstein's (1967, II, iv) claim that we can only know another person's soul through their body.

3. On the importance of both psychological continuity and bodily continuity for personal identity, and specifically, as features of a person's biography, see David Wiggins (1979).

4. Emmanuel Levinas (2000) and Françoise Dastur (1996, pp. 42–9) point to the role that the death of others has in providing us with knowledge of death. In so doing, they draw upon but are critical of Heidegger's assessment (1962, pp. 281–5) of the role of the death of the other in revealing that only Dasein can die its own death.

5. The idea that more of a good is better than less, other things being equal, is promoted by Nagel (1970, p. 2), Williams (1973b, p. 84) and Jonathan Glover (1977, p. 55).

6. As was noted in Chapter 1, Williams observes that it is testament to the success of our activities and attachments, and hence our categorical desires, that the question of the meaning of our lives does not arise.

7. Frederick Kaufman (1999) maintains that all successful accounts of the badness of death are forms of the deprivation thesis.

8. The ontological problem is sometimes called the problem of the subject. I am grateful to Suzanne Uniacke for pointing out that one way of clarifying the distinction between the two Epicurean problems of non-existence is to describe them as phenomenological and ontological.

9. John Goodwin's notes to Lucretius make this point about infinity clear (Lucretius, 1994, p. 221, n. 113).

10. Stephen Rosenbaum (1989) maintains that to interpret Epicurus as claiming that it does not matter when we die is to misunderstand him. What Epicurus sought was to undermine the fear of death by observing that it involves non-existence and hence no good or bad experiences.

11. Gerald Gruman (1966, p. 14) suggests that one reason for the Epicurean view is that during his lifetime there was no appreciation of the control that could be obtained over nature or of the progress that might be made concerning human activities that would make prolonging life desirable. Gruman considers other arguments to be found in myth, religion and works of ancient philosophers that oppose seeking greater longevity, which he labels 'Apologist'. Christine Overall (2003, pp. 23–63) provides examples of more recent arguments against increasing life spans; Martien Pijnenburg and Carlo Leget (2007) provide an argument against extending life spans based on its consequences.

12. The tension between the subjective and objective views is a continuing theme in Nagel's work (1979b, 1986, 1991). Nagel argues that neither view can supersede the other: the objective view cannot account for the personal nature of the subjective view, while the objective point of view is never truly objective.

13. Nagel (1970, p. 6) queries whether the intelligent adult can be said to continue to exist after the accident.

14. This is complicated further by accounts of personal identity that use the living body as the criterion for personal identity, such as those of Olson (1997) and DeGrazia (2005).

15. A benefit of Nagel's argument is that it explains what is wrong with Lucretius' claim that it does not matter when we die. If death were of finite duration, to die later rather than earlier entails enduring less of death, thereby making it better to die later than earlier. The infinite duration of death, however, means those who die later will endure the same, infinite duration of death as those who die earlier. Yet, if it were the case that the

duration of death could determine whether it is better to die earlier or later, this must be because the duration of death is experiential, which contradicts both the Epicurean and Lucretian views and the deprivation thesis (Williams, 1973b, p. 84). What makes death bad is the loss of life, and this implies that to die earlier is worse than dying later, a point that becomes clearer as the argument progresses.

16. Similar arguments are made by Ben Bradley (2004), Neil Feit (2002), William Grey (1999) and Geoffrey Scarre (2007).
17. The average life expectancy for Shakespeare and Proust would have been less than 80 years.
18. The example of Keats and Tolstoy is from Nagel (1970, p. 9). Nagel cites Keats's age as being 24 when he died.
19. McMahan (1988, p. 45; 2002, pp. 113–14) relies upon David Lewis's description of the truth values of counterfactuals to identify the nearest possible world in which the immediate causes of X's death are missing as the one in which the antecedent, 'if X had not died', is true.
20. The example of Joe is from McMahan (1988, pp. 42–9); he provides a more detailed discussion of these issues in his later work (McMahan, 2002).
21. This point is made by Belshaw (2009, p. 109) against McMahan's critique of the deprivation thesis. Nagel's account (1970) of the misfortune of death does seem to acknowledge this, and his atemporal account of the timing of death is consistent with his view that death is always bad.
22. McMahan (2002, pp. 124–7) is sceptical about the role the normal length of life can play in determining the misfortune of death.
23. Richard Wollheim (1984, pp. 265–6) argues that an objection to immortality is that the length of life would remove our present reasons for choosing certain of our activities.
24. This argument is made by Brueckner and Fischer (1986, 1993a, 1993b).
25. Nagel (1970, p. 8, n. 3) considers, albeit briefly, the possibility that someone could have been born earlier while retaining their genetic identity, drawing on an example by Robert Nozick, and concludes that this possibility suggests that the solution to the asymmetry problem depends upon our past and future concerns about our own lives.
26. Saul Kripke (1981, p. 62) argues that we do not need a precise date of birth for someone to be identified as the same person, so long as their origins remain the same. It should be noted that Kripke is not providing a theory of personal identity, although his ideas about origins have been used in this way.
27. In Karel Čapek's play, Elina Makropulos is 337 years old.
28. McMahan (2002, p. 100) rejects Williams's argument and maintains that it is rational for an individual now to have egoistic concerns about a future individual without them being one and the same individual.

3 Justifying the Means

1. R. G. Frey (1996) recognises, in the context of animal experimentation, that medicine becomes compromised by the use of immoral means.
2. An issue I will not consider is whether it would be morally permissible to use or to continue using information that was wrongfully obtained. For example, Nazi

experiments on hypothermia were wrongfully conducted but the data collected is recognised by some as being scientifically useful (Moe, 1984; Post, 1991).

3. Two notable, recent attempts to show that animals cannot suffer are by Peter Harrison (1991) and Peter Carruthers (1992).

4. Peter Singer (1995) provides a wide range of empirical evidence with the purpose of questioning the efficacy of animal experimentation.

5. Tom Regan provides a detailed criticism of preference utilitarianism, the version of utilitarianism supported by Singer (Cohen and Regan, 2001, pp. 180–8).

6. Williams (2006) provides a fuller account of his view on the 'human prejudice'.

7. Singer once used the example of a 'moral ledger' for calculating the aggregate of preference satisfaction, such as the ledgers used by accountants, which was criticised by H. L. A. Hart (Singer, 1993b, p. 128).

8. The claim that morality is 'about' human beings raises questions regarding the possibility of aliens with similar intellectual and physical capacities to human moral agents. How we respond to aliens depends upon their morally salient features, as will become clear in later chapters. Rosalind Hursthouse (1987, pp. 247–55) and Williams (2006, pp. 148–52) raise and respond to the issue of aliens.

9. The example of Jim and the Indians is similar in many respects to Philippa Foot's (1967) example of the trolley problem. It is not obvious that Jim should probably kill one of the Indians. G. E. M. Anscombe (1967) and John Taurek (1977) argue that the numbers involved in such conflicts do not count.

10. Williams's comments during his discussion of integrity suggest he is aware of the problem I raise.

11. Uniacke (2004a) discusses the role of the context of an act that harms, and so may cause suffering, in determining whether it is wrong.

12. The Nuffield Council on Bioethics (2005, p. 7) notes that the precise numbers of animals used are difficult to obtain but are estimated to be between 50 and 100 million.

13. Søren Holm (2002) provides an overview of the broad range of ethical issues involved in the stem cell debate, as does the Chief Medical Officer's Report (Department of Health, 2000).

14. Jenny Teichman (1985) provides an overview of the many different understandings of the term 'person' in philosophy; Michael Tooley (1998a) and Mary Anne Warren (1997) discuss the range of psychological characteristics that are often associated with Personhood.

15. H. G. Woodger's observation is also used by Holland (1990).

16. Jason Eberl (2000) describes the potential Person and Persons as unified individuals with ongoing identity, which are ontologically unique.

17. The problems of identity associated with the early embryo are well known in logic. Graham Priest (2000, pp. 68–9) discusses them in relation to amoebas and in so doing uses the example of either X or Y ceasing to exist the moment after division occurs.

18. This was the conclusion of the Warnock Committee (Warnock, 1985), which set out the principles contained in the Human Fertilisation and Embryology Act of 1990.

19. The debate about using spare embryos and creating embryos for medical purposes is discussed by Nicole Gerrand (1993) and K. Devolder (2005).

4 Longevity and the Problem of Overpopulation

1. The cultural background will define the ideal of the good life for a given society, which will determine the nominal optimal population level. Within each clearing will be sub-cultures that identify different ways for pursuing this ideal of the good life.
2. Typical examples are Glover (1977) and Derek Parfit (1984).
3. For this reason, as John Broome (2005) argues, the addition of people to a population is never value-neutral, a position supported by Jan Narveson (1973).
4. What constitutes deliberately having a child is contentious and is assessed by David Archard (2004).
5. Timothy Sprigge (1968, pp. 337–8) observes that it is possible to reverse Narveson's argument so that the actual person whom our actions affect, and benefit, is one whose life is worthwhile, but we do not prevent a person suffering if we fail to bring them into existence. McMahan (1981, pp. 99–109) offers an overview of the difficulties with Narveson's argument.
6. Tooley (1998b) provides a more detailed argument in favour of the asymmetry argument based on distinguishing between evaluative and normative properties. Broome (2005) also makes a similar distinction when he claims that bringing a person into existence is not value-neutral, but the normative requirements about so doing may be.
7. Gustaf Arrhenius (2008) provides a recent assessment of the dilemma between having more lives or longer lives.
8. My account of Parfit's Mere Addition Paradox draws on his shorter explication of it (1986). A more detailed version is to be found in *Reasons and Persons* (1984), but this lacks his account of Perfectionism.
9. Alan Carter (1999, p. 300) makes this point. John Rawls argues that his two principles of justice are a special case of a more general conception of justice, which maintains that '[a]ll social values – liberty and opportunity, income and wealth, and the bases of self-respect – are to be distributed equally unless an unequal distribution of any, or all, of these values is to everyone's advantage' (1971, p. 62).
10. Carter (1999, pp. 296–7) refers to an example by Singer of the cost of the construction of the Sydney Opera House, when the then Australian government donated the equivalent of a twelfth of this sum to the Bengal refugee crisis.
11. To be fair to Parfit, the new Principle of Beneficence that he searches for, his Theory X, seeks more than to find a solution to the Repugnant Conclusion.
12. John Harris (2000) also points to the possibility of overpopulation and the potential consequences for having children, although he does not consider the implications to the extent outlined here. David Gems (2003) recognises the threat of overpopulation from increasing life spans, but is sceptical about it leading to overpopulation given the present decline of fertility in the West.

5 Ending Lives

1. Harris (1985, pp. 91–4) provides an outline of the argument, which is developed by Alan Williams (1997).
2. Harris (1985, p. 91) claims that to fail to live for the full fair innings is to suffer an injustice, although in what way it is an injustice is unclear.
3. Norman Daniels (2008, pp. 171–81) provides a lifetime account of healthcare distribution, which I consider in the following chapter.
4. Wiggins (1987) provides a thorough examination of this connection. Amartya Sen (1980) grounds the right to healthcare in capabilities rather than needs, but, as Daniels (2008, p. 78) observes, the general line of argument is similar to that of a needs-based approach.
5. What, if any healthcare would be available to those who live longer than the fair innings is open to question, which is why I refer to the 'broad range' of healthcare options, assuming that some healthcare, even if only palliative care, would be available. I will return to this issue later in my discussion.
6. To be clear, in considering what a policy limiting longevity might involve, I do not advocate restricting life spans. I am simply considering the implications of such a policy.
7. In making this assumption, I am discounting two possibilities: first, that the economic resources do not exist to develop the biomedicines necessary for increasing life spans; and, second, given the argument that there is no natural life span, and that there may be a need to restrict access to healthcare and other resources, a policy might be established that reduces life spans to below present levels, thereby releasing resources. How this might be achieved is not entirely clear, but it is a logical possibility given my argument about the relationship between how long we can live and human activities.
8. John Hardwig (2000, p. 129) does not restrict the duty to die to the elderly, but maintains that it becomes greater as one grows older and has experienced more of life than one's family. Margaret Battin (2000) outlines a justification for a duty to die for reasons similar to those considered here, but she also proposes that a duty to die might, with the appropriate social institutions, be used to equalise life expectancies throughout the world. Overall (2003) provides a broad critique of the duty to die as an argument against prolonging life, raising different objections from those that I focus upon.
9. With this observation about the duty to die, I adapt a criticism by Anscombe (1967) of an example by Foot (1967) in her critique of the Doctrine of Double Effect.
10. The QALY has been accused of being an ageist means for distributing healthcare (Harris, 1987).
11. Harris (2000), and Robert Garland (1990, pp. 244–5) in relation to Herodotus, refer to the policy of killing people in order to distribute resources fairly as enforced euthanasia, but to do so is incorrect, as will become clear.
12. Hugo Bedau (1968) outlines the many different claims that the concept of the right to life has been thought to encompass, such as the right to various forms of welfare support.
13. John Keown (2002) and Richard Tur (2002) provide various definitions of euthanasia. Keown also describes suicide and assisted suicide.

14. There is some dispute about whether there is a substantive moral difference between assisted suicide and voluntary euthanasia (Keown, 2002, pp. 31–6). My assumption should not be taken to imply that there is no substantive difference; I maintain only that once the arguments are in place to justify suicide and voluntary euthanasia, then there is nothing more of any moral substance to add in order to justify assisted suicide.

15. In describing the sanctity of life doctrine in this way, I draw on David Oderberg's definition: '[i]t is always and everywhere a grave moral wrong intentionally to take the life of an innocent human being' (2000b, p. 147).

16. For a critique of Ronald Dworkin's secular (1994) account of the sanctity of life, see Uniacke (2004b).

17. Dworkin (1994, p. 85) and Uniacke (2004b) also observe that it is important to distinguish between the value of life and the conventions governing the ending of life.

18. An alternative explanation for why homicide in self-defence and just warfare may be permissible – and, in contrast to the Doctrine of Double Effect, intentional – is that they involve unjust conduct. Uniacke (1994) discusses the role of conduct in permissible homicide in self-defence.

19. W. D. Ross (1930, pp. 48–55) outlines the relationship between four logically independent statements connecting rights with duties. The statements are (p. 48):

    1. A right of A against B implies a duty of B to A.
    2. A duty of B to A implies a right of A against B.
    3. A right of A against B implies a duty of A to B.
    4. A duty of A to B implies a right of A against B.

20. Brenda Almond (1993) provides an account of rights as abstract nouns of a similar nature to duties and obligations. Ross (1930, pp. 50–2) appears to identify rights as markers delineating the nature of human powers and relationships, but I draw on the work of John Finnis (1977, p. 219) who describes them as markers.

21. Indeed, this is the reason why Feinberg discusses the three methods for resolving rights conflict: to consider possible ways of conceiving of the right to life as being absolute in some sense while also permitting killing in certain circumstances. On the basis of this analysis, Feinberg (1978, p. 104) denies that the right to life is absolute, but he instead claims it to be inalienable.

22. Wiggins (1998, p. 272) suggests Susan Hurley first used the term *pro tanto* in place of *prima facie* for the reasons cited.

23. Feinberg (1978, pp. 98–9) refers to William Frankena's (1955) description of *prima facie* rights, but I draw on Ross (1930) and Dancy's (1993a; 1993b, pp. 93–6) explication of Ross's ideas.

24. For example, Aristotle claims, 'for nothing perceptible is easily defined, and [since] these [circumstances of virtuous and vicious action] are particulars, the judgement about them depends upon perception' (1985, 1109b20–4, text in brackets is from the original translation). Wiggins (1975) provides a detailed analysis of Aristotle's account of practical reason.

25. Finnis (1977) describes the process of reasoning in natural law as practicable reasonableness. A. P. D'Entrèves (1970) describes the role of reason in determining natural justice. The natural law account of justice is influential for my understanding of resolving moral conflict.
26. Heidegger (1962, pp. 287–8) draws an analogy between the time of death and a ripening fruit. A fruit can die before it ripens, and so many people will die while their activities and attachments still make their lives fulfilling. A fruit can also become overripe, where a person might outlive their activities, as described, or experience the tedium of immortality.
27. Foot (1977) describes euthanasia as involving a conflict between the virtues of justice and charity.
28. There are a number of different interpretations of the perfect and imperfect distinction between duties. George Rainbolt (2000) identifies eight. I also draw on Onora O'Neill's description (1996, pp. 81–3, 86–8).
29. I have not addressed practical concerns about voluntary euthanasia leading to involuntary euthanasia. It is important to ensure that euthanasia is voluntary, and this will involve a number of difficulties, but sufficient safeguards can be put in place to prevent a slide from voluntary to involuntary euthanasia.

6   Partiality and Equality

1. The earlier discussion of using human embryos in medicine did not consider their right to life. It should become clear as the discussion develops why the embryo lacks a right to life.
2. The issue of intergenerational justice is often described as a conflict between the needs of the young and of the old, as is found with the Prudential Life Span Account, but this leaves open whether a 'middle' generation might have different needs.
3. The appeal and difficulty of explaining and uniting the idea of equality with claims for equal treatment of people are discussed by Isaiah Berlin (1956) and Wollheim (1956).
4. Nagel (1991) contrasts the personal perspective with the impersonal perspective, which I understand to be analogous to his earlier contrasting of the subjective and objective distinction (1979b, 1986). I retain the term 'subjective' to describe an individual's perspective on the world. The objective perspective can be either personal or impersonal. In the discussion of death, the objective perspective on the deceased is personal because it focuses on a particular individual. In the case of equality, the objective perspective is impersonal because it does not centre on a particular individual.
5. For my description of a person's character, I draw on Aristotle (1985) and Urmson's commentary (1988), although I have a broader view of the nature of character than Aristotle.
6. The elucidation of the partialist's critique of impartialism largely draws on Susan Wolf (1982). 'Morally beneficial' is Wolf's term (p. 422).
7. Whether or not acts can be supererogatory is greatly disputed. The debate between Elizabeth Pybus (1982), who denies their possibility, Patricia McGoldrick (1984) and Russell Jacobs (1987), who support supererogation,

albeit to varying degrees, and Francis Kamm's (1985) discussion, offers an insight into the complex issues that it raises.

8. Cottingham (1996, pp. 68–70) observes that issues of justice prove difficult for partialism, but I propose a more constrained form of partialist concern, which, as I will argue, can take account of justice and fairness.

9. Cottingham refers only to what Williams (1976a) and Nagel (1976) identify as 'constitutive' luck.

10. Raz (2001, p. 126) describes the account of respect for people that I propose as the by-product view of respect, an account to which he once 'tended to lean'. The by-product view denies that there is a distinctive duty of respect; we respect people, on this view, when we treat them according to the fundamental moral requirements.

11. Williams (1962, pp. 240–1) points to the relevance of reasons for or against claims of equality.

12. McMahan (2002, pp. 233–65) provides an alternative account of equal respect.

13. The account I provide is merely a brief sketch of the meta-ethical theory, which I consider best explains how we can know how to behave morally in a given situation, but it should provide a good idea of the approach I take.

14. Both H. A. Prichard (1912, p. 7) and Ross (1930, p. 20, n.1) describe the immediacy of perceiving the saliency of a moral property, although only Prichard compares it to the immediacy of mathematics. Williams (1988a, pp. 182–5) draws a distinction between perceptual and mathematical intuitionism.

15. I borrow Williams's use (1980) of φ for verbs of action and extend this to examples of the non-specific moral properties of a situation.

16. I draw on Dancy's (1993b, pp. 93–6) elucidation of Ross.

17. I draw primarily on Dancy (1993b, chs 4 and 7). Dancy (2004) provides a more recent and fuller statement of his account of particularism.

18. Dancy (1993b, pp. 109–11) is responding to Williams's description (1965) of agent regret.

19. Wiggins (1975) interprets Aristotle as a particularist about rules, of which Roger Crisp (2000) is critical.

20. Dancy (1993b, pp. 4–5) points out the problem of evil, but also the difficulties that accidie and amoralism cause intuitionist theories of moral epistemology.

21. The example of kindness is John McDowell's (1979, pp. 332–3). Williams (1993, p. 217, n. 7) maintains that the link between being able to recognise an evaluative concept and being predisposed towards it is a Wittgensteinian idea, first recognised by Philippa Foot and Iris Murdoch in the 1950s.

22. Cottingham (1996, p. 69) maintains that the patricians of the Roman Empire were able to live the good life despite the widespread injustices that existed beyond their social sphere.

23. Anscombe (1958) and Alasdair MacIntyre (1985), for example, view the moral virtues and moral duties and obligations as, in O'Neill's phrase (1996), 'antithetical'.

24. Cottingham (1996, pp. 68–9), for example, suggests that the virtues cannot provide an adequate account of justice. This is in part because the virtues are associated with partialist ethics, where justice implies impartialism (although this is not a difficulty for my account), but also because the virtue of justice can be neither excessive nor deficient, unlike the other cardinal virtues.

Bibliography

Alderman, H. (1978). 'Heidegger's Critique of Science and Technology', in Michael Murray (ed.). *Heidegger and Modern Philosophy* (New Haven: Yale University Press).

Almond, B. (1993). 'Rights', in Singer (1993a).

Almond, B. (1999). 'Biomedical Technology in a Humanistic Culture', *Public Affairs Quarterly*, 13, 229–40.

Anscombe, G. E. M. (1958). 'Modern Moral Philosophy', *Philosophy*, 33, 1–19.

Anscombe, G. E. M. (1967). 'Who is Wronged?', *The Oxford Review*, 5, 16–17.

Archard, D. (2004). 'Wrongful Life', *Philosophy*, 79, 403–20.

Aristotle (1985). *Nichomachean Ethics*, Terence Irwin (trans.) (Indianapolis: Hackett).

Aristotle (2004). *Metaphysics*, Hugh Lawson-Tancred (trans.) (London: Penguin).

Arluke, A. (1992). 'Trapped in a Guilt Cage', *New Scientist*, 134 (1815), 33–5.

Arrhenius, G. (2008). 'Life Extension versus Replacement', *Journal of Applied Philosophy*, 25, 211–27

Austad, S. N. (1997). *Why We Age. What Science is Discovering about the Body's Journey through Life* (New York: John Wiley & Sons Inc.).

Baird, R. M. and Rosenbaum, S. E. (eds) (1991). *Animal Experimentation. The Moral Issues* (Amherst, N.Y.: Prometheus Books).

Battin, M. P. (1998). 'Population Issues', in Kuhse and Singer (1998).

Battin, M. P. (2000). 'Global Life Expectancies and the Duty to Die', in Humber and Almeder (2000).

Bauman, Z. (1992). *Mortality, Immortality and Other Life Strategies* (Stanford, Calif.: Stanford University Press).

Becker, L. (1975). 'Human Beings: The Boundaries of the Concept', *Philosophy and Public Affairs*, 4, 334–59.

Bedau, H. (1968). 'The Right to Life', *The Monist*, 52, 550–72.

Belshaw, C. (2009). *Annihilation* (Stocksfield: Acumen).

Berlin, I. (1956). 'Equality', *Proceedings of the Aristotelian Society*, 56, 301–26.

Blackford, R (2006). 'Sinning Against Nature: The Theory of Background Conditions', *Journal of Medical Ethics*, 32, 629–34.

Boorse, C. (1975). 'On the Distinction between Disease and Illness', *Philosophy and Public Affairs*, 5, 49–68.

Boorse, C. (1977). 'Health as a Theoretical Concept', *Philosophy of Science*, 44, 542–73.

Bradley, B. (2004). 'When is Death Bad for the One Who Dies', *Noûs*, 38, 1–28.

Broome, J. (2005). 'Should We Value Population?', *Journal of Political Philosophy*, 13, 399–413.

Brueckner, A. and Fischer, J. M. (1986). 'Why is Death Bad?', *Philosophical Studies*, 50, 213–21.

Brueckner, A. and Fischer, J.M. (1993a). 'The Asymmetry of Early Death and Late Birth', *Philosophical Studies*, 71, 327–31.

Brueckner, A. Fischer, J. M. (1993b). 'Death's Badness', *Pacific Philosophical Quarterly*, 74, 37–45.

Camus, A. (1955). *The Myth of Sisyphus*, Justin O'Brien (trans.) (Harmondsworth: Penguin).

Čapek, K. (1922). 'The Makropulos Case', Peter Majer and Cathy Porter (trans.) in Čapek, *Four Plays* (London: Methuen, 1999).

Carnes, B. A., Olshansky, S. J. and Grahn, D. (2003). 'Biological Evidence for the Limits to the Duration of Life', *Biogerontology*, 4, 31–4.

Carruthers, P. (1992). *The Animals Issue. Moral Theory in Practice* (Cambridge: Cambridge University Press).

Carter, A. (1999). 'Moral Theory and Global Population', *Proceedings of the Aristotelian Society*, 99, 289–313.

Coale, A. (1964). 'How a Population Ages and Grows Young', reprinted in Kammyer (1975).

Cohen, C. and Regan, T. (2001). *The Animal Rights Debate* (Oxford: Rowman and Littlefield).

Cole-Turner, R. (1998). 'Do Means Matter', in Parens (1998).

Cottingham, J. (1996). 'Partiality and the Virtues', in Crisp (1996).

Cottingham, J. (1997). 'The Ethical Credentials of Partiality', *Proceedings of the Aristotelian Society*, 98, 1–21.

Cottingham, J. (1998). *Philosophy and the Good Life. Reason and the Passions in Greek, Cartesian and Psychoanalytic Ethics* (Cambridge: Cambridge University Press).

Crisp, R (1996). *How Should One Live? Essays on the Virtues* (Oxford: Oxford University Press).

Crisp, R. (ed.) (2000). 'Particularizing Particularism', in Brad Hooker and Margaret Little (eds). *Moral Particularism* (Oxford: Clarendon Press).

D'Entrèves, A. P. (1970). *Natural Law*, 2nd edn (London: Hutchinson).

Dancy, J. (1983). 'Ethical Particularism and Morally Relevant Properties', *Mind*, 92, 530–47.

Dancy, J. (1993a). 'An Ethic of *Prima Facie* Duties', in Singer (1993a).

Dancy, J. (1993b). *Moral Reasons* (Oxford: Blackwell).

Dancy, J. (2004). *Ethics without Principles* (Oxford: Clarendon Press).

Daniels, N. (2008). *Just Health. Meeting Health Needs Fairly* (Cambridge: Cambridge University Press).

Dastur, F. (1996). *Death. An Essay on Finitude*, John Llewelyn (trans.) (London: Althone).

Davidson, D. (1974). 'On the Very Idea of a Conceptual Scheme', reprinted in Davidson, *Inquiries into Truth and Interpretation* (Oxford: Clarendon Press, 1984).

Deckers, J. (2007). 'Why Eberl is Wrong: Reflections on the Beginning of Personhood', *Bioethics*, 21, 270–82.

DeGrazia, D. (2005). *Human Identity and Bioethics* (Cambridge: Cambridge University Press).

Department of Health (2000). *Stem Cell Research: Medical Progress with Responsibility. A Report From the Chief Medical Officer's Expert Group Reviewing the Potential of Developments in Stem Cell Research and Cell Nuclear Replacement to Benefit Human Health* (London)

Devolder, K. (2005). 'Creating and Sacrificing Embryos for Stem Cells', *Journal of Medical Ethics*, 31, 366–70.

Dreyfus, H. (1991). *Being-in-the-World. A Commentary on Heidegger's* Being and Time, *Division I* (Cambridge, Mass.: MIT Press).

Dreyfus, H. (1993). 'Heidegger on the Connection between Nihilism, Art, Technology, and Politics', in Charles Guignon (ed.). *The Cambridge Companion to Heidegger* (Cambridge: Cambridge University Press).

Dworkin, R. (1994). *Life's Dominion. An Argument about Abortion, Euthanasia and Individual Freedom* (New York: Vintage).

Eberl, J. T. (2000). 'The Beginning of Personhood: A Thomistic Biological Analysis', *Bioethics*, 14, 134–57.

Eliot, T. S. (1962) *Notes towards the Definition of Culture* (London: Faber and Faber).

Elliott, C. (1998). 'The Tyranny of Happiness: Ethics and Cosmetic Psychopharmacology', in Parens (1998).

Epicurus (1964). 'Letter to Menoeceus', in Russell M. Geer (ed. and trans.). *Letters, Principle Doctrines, and Vatican Sayings* (Indianapolis: Bobbs-Merrill Company).

Feinberg, J. (1978). 'Voluntary Euthanasia and the Inalienable Right to Life', *Philosophy and Public Affairs*, 7, 93–123.

Feit, N. (2002). 'The Time of Death's Misfortune', *Noûs*, 36, 359–83.

Finnis, J. (1977). *Natural Law and Natural Rights* (Oxford: Clarendon Press).

Foot, P. (1967). 'The Problem of Abortion and the Doctrine of Double Effect', reprinted in Foot (1978).

Foot, P. (1977). 'Euthanasia', reprinted in Foot (1978),

Foot, P. (1978). *Virtues and Vices* (Oxford: Blackwell).

Frankena, W. (1955). 'Natural and Inalienable Rights', *Philosophical Review*, 64, 212–32.

Freud, S. (1920). *Beyond the Pleasure Principle*, reprinted in Freud (1991).

Freud, S. (1923). *The Ego and the Id*, reprinted in Freud (1991).

Freud, S. (1991). *On Metapsychology: The Theory of Psychoanalysis. Penguin Freud Library*, 11, James Strachey (trans.) and Angela Richards (ed.) (Harmondsworth: Penguin).

Frey, R. G. (1996). 'Medicine, Animal Experimentation and the Moral Problem of Unfortunate Humans', *Social Policy and Philosophy*, 13, 181–211.

Fukuyama, F. (2002). *Our Posthuman Future* (London: Polity).

Garland, R. (1990). *The Greek Way of Life* (London: Duckworth).

Gems, D. (2003). 'Is More Life Always Better? The New Biology of Aging and the Meaning of Life', *Hastings Center Report*, 33, 31–9.

Gerrand, N. (1993). 'Creating Embryos for Research', *Journal of Applied Philosophy*, 10, 175–87.

Glover, J. (1977). *Causing Death and Saving Lives* (Harmondsworth: Penguin).

Gorer, G. (1955). 'The Pornography of Death', *Encounter*, 5, 49–52.

Grey, W. (1999). 'Epicurus and the Harm of Death', *Australasian Journal of Philosophy*, 77, 358–64.

Gruman, G. J. (1966). 'A History of Ideas about the Prolongation of Life: The Evolution of Prolongevity Hypotheses to 1800', *Transactions of the American Philosophical Society*, 56 (9), 1–102.

Guarente, L. and Kenyon, C. (2000). 'Genetic Pathways that Regulate Ageing in Model Organisms', *Nature*, 408, 255–62.

Guillebaud, J. and Hayes, P. (2008). 'Editorial: Population Growth and Climate Change', *British Medical Journal*, 337, a576.

Haber, C. (2004). 'Life Extension and History: The Continual Search for the Fountain of Youth', *Journal of Gerontology*, 59A (6), 515–22.

Hadley, E. C., Lakatta, E. G., Morrison-Bogorad, M., Warner, H. R. and Hodes, R. J. (2005). 'The Future of Aging Therapies', *Cell*, 120, 557–67.

Hardwig, J. (2000). *Is There a Duty to Die? And Other Essays in Bioethics* (London: Routledge).

Hare, R. M. (1988). 'Possible People', reprinted in Hare, *Essays on Bioethics* (Oxford: Clarendon, 1993).

Harris, J. (1985). *The Value of Life* (London: Routledge).

Harris, J. (1987). 'QALYfying the Value of Life', *Journal of Medical Ethics*, 13, 117–23.

Harris, J. (2000). 'Intimations of Immortality', *Science*, 288, 59.

Harrison, P. (1991). 'Do Animals Feel Pain?', *Philosophy*, 66, 25–40.

Hart, H. L. A (1967). 'Punishment and Intention', reprinted in Hart, *Punishment and Responsibility* (Oxford: Clarendon, 1968).

Hayflick, L. (2000). 'The Future of Ageing', *Nature*, 408, 267–9.

Heidegger, M. (1962). *Being and Time*, John Macquarrie and Edward Robinson (trans.) (Oxford: Blackwell).

Heidegger, M. (1967). 'Modern Science, Metaphysics and Mathematics', W. B. Barton Jr and Vera Deutsch (trans.), reprinted in Heidegger (1993).

Heidegger, M. (1971). 'The Origin of the Work of Art', Albert Hofstadter (trans.), reprinted in Heidegger (1993)

Heidegger, M. (1977). 'The Question Concerning Technology', William Lovitt (trans.), reprinted in Heidegger (1993).

Heidegger, M. (1993). *Basic Writings*, David Farrell Krell (ed.) 2nd edn (London: Rouledge).

Herodotus (1997). *The Histories*, George Rawlinson (trans.) and David Campbell (rev.) (London: Everyman).

Holland, A. (1990). 'A Fortnight of My Life is Missing: A Discussion of the Status of the Human "Pre-Embryo"', *Journal of Applied Philosophy*, 7, 25–37.

Holm, S. (2002). 'Going to the Roots of the Stem Cell Controversy', *Bioethics*, 16, 493–507.

Humber, J. A. and Almeder, R. F. (2000). *Is There a Duty to Die?* (Totowa, NJ: Humana Press).

Hursthouse, R. (1987). *Beginning Lives* (Oxford: Blackwell/Open University).

Ihde, D. (1979). 'Heidegger's Philosophy of Technology', reprinted in Robert C. Scharff and Val Dusek (eds). *Philosophy of Technology: The Technological Condition. An Anthology* (Oxford: Blackwell, 2003).

Inwood, M. (1997). *Heidegger* (Oxford: Oxford University Press).

Jacobs, R. A. (1987). 'Obligation, Supererogation and Self-Sacrifice', *Philosophy*, 62, 96–101.

Jaenisch, R. (2004). 'Human Cloning: The Science and Ethics of Nuclear Transportation', *New England Journal of Medicine*, 351, 2787–91.

Johnstone, H.W. (1976). 'Sleep and Death', *The Monist*, 59, 218–33.

Jonas, H. (1979). 'Towards a Philosophy of Technology', *Hastings Center Report*, 9 (1), 34–43.

Juengst, E. T. (1998). 'What Does Enhancement Mean?' in Parens (1998).

Jeungst, E. T., Binstock, R. H., Mehlman, M., Post, S. G. and Whitehouse, P. (2003). 'Biogerontology, "Anti-aging Medicine", and the Challenges of Human Enhancement', *Hastings Center Report*, 33, 21–30

Kamm, F. M. (1985). 'Supererogation and Obligation', *Journal of Philosophy*, 82, 118–38.

Kammeyer, K. C. W. (ed.) (1975). *Population Studies. Selected Essays and Research*, 2nd edn (Chicago: Rand McNally College Publishing Company).

Kaufman, F. (1996). 'Death and Deprivation: Or, Why Lucretius' Symmetry Argument Fails', *Australasian Journal of Philosophy*, 74, 305–12.

Kaufman, F. (1999). 'Pre-Vital and Post-Mortem Non-Existence', *American Philosophical Quarterly*, 36, 1–19.

Kenyon, C. (1996). 'Ponce d'elegans: Genetic Quest for the Fountain of Youth', *Cell*, 84, 501–4.

Keown, J. (2002). *Euthanasia, Ethics and Public Policy. An Argument Against Legislation* (Cambridge: Cambridge University Press).

King, C. Daly (1945). 'The Meaning of Normal', *Yale Journal of Biology and Medicine*, 17, 493–501.

Kirkwood, T. B. L. (1977). 'Evolution of Ageing', *Nature*, 270, 301–4.

Kirkwood, T. B. L. (1997). 'The Origins of Human Ageing', *Philosophical Transactions of the Royal Society of London. Series B, Biological Sciences*, 352, 1765–72.

Kirkwood, T. B. L. (2000). *The Time of Our Lives. The Science of Human Ageing* (London: Phoenix).

Kirkwood, T. B. L. (2005). 'Understanding the Odd Science of Ageing', *Cell*, 120, 437–47.

Kirkwood, T. B. L. (2008). 'A Systematic Look at an Age Old Problem', *Nature*, 451, 644–7.

Kirkwood, T. B. L. and Austad, S. N. (2000). 'Why Do We Age?', *Nature*, 408, 233–8.

Kirkwood, T. B. L. and Cremer, T. (1982). 'Cytogerontology Since 1881: A Reappraisal of August Weismann and a Review of Modern Progress', *Human Genetics*, 60, 101–21.

Kirkwood, T. B. L. and Rose, M. R. (1991). 'Evolution of Senescence: Late Survival Sacrificed for Reproduction', *Philosophical Transactions of the Royal Society of London. Series B, Biological Sciences*, 332, 15–24.

Kripke, S. (1981). *Naming and Necessity* (Oxford: Blackwell).

Kuhn, T. S. (1996). *The Structure of Scientific Revolutions*, 3rd edn (London: Chicago University Press).

Kuhse, H. and Singer, P. (eds) (1998). *A Companion to Bioethics* (Oxford: Blackwell).

LaFollette, H. (ed) (1997). *Ethics in Practic. An Anthology* (Oxford: Blackwell).

Levinas, E. (2000). *God, Death and Time*, Bettina Bergo (trans.) (Stanford, Calif: Stanford University Press).

Little, M. O. (1998). 'Cosmetic Surgery, Suspect Norms, and the Ethics of Complicity', in Parens (1998).

Lucretius (1994). *On the Nature of the Universe*, R. E. Latham (trans.) and John Goodwin (rev.) (London: Penguin).

Luper-Foy, S. (1987). 'Annihilation', *Philosophical Quarterly*, 37, 233–52.

McDowell, J. (1979). 'Virtue and Reason', *The Monist*, 62, 331–50.

McGoldrick, P. M. (1984). 'Saints and Heroes: A Plea for the Supererogatory', *Philosophy*, 59, 523–8.

MacIntyre, A. (1985). *After Virtue. A Study in Moral Theory*, 2nd edn (London: Duckworth).

McKenny, G. P. (1998). 'Enhancements and the Ethical Significance of Vulnerability', in Parens (1998).

Mackie, J. L. (1977). *Ethics. Inventing Right and Wrong* (Harmondsworth: Penguin).

McMahan, J. (1981). 'Problems of Population Theory', *Ethics*, 92, 96–127.

McMahan, J. (1988). 'Death and the Value of Life', *Ethics*, 99, 32–61

McMahan, J. (2002). *The Ethics of Killing. Ethics at the Margins of Life* (Oxford: Oxford University Press).

Malthus, T. (1798). *First Essays on the Principle of Population* (London: Macmillan).

Marquis, D. (1989). 'Why Abortion is Immoral', *Journal of Philosophy*, 86, 183–202.

Marquis, D. (1997). 'An Argument that Abortion is Wrong', in LaFollette (1997).

Maynard Smith, J. (1962). 'Review Lectures on Senescence. I: The Causes of Ageing', *Proceeding of the Royal Society of London. Series B, Biological Sciences*, 157, 115–27.

Medawar, P. B. (1952). *An Unsolved Problem of Biology* (London: H.K. Lewis & Co.).

Meissner, A. and Jaenisch, R. (2006). 'Generation of Nuclear Transfer-Derived Pluripotent ES Cells from Cloned Cdx-2 Deficient Blastocysts', *Nature*, 439, 213–5.

Moe, K. (1984). 'Should the Nazis Research Data be Cited?' *Hastings Center Report*, 14, 5–7.

Mulhall, S. (1996). *Heidegger and Being and Time* (London and New York: Routledge).

Nagel, T. (1970). 'Death', reprinted in Nagel (1979a).

Nagel, T. (1971). 'The Absurd', reprinted in Nagel (1979a).

Nagel, T. (1976). 'Moral Luck', reprinted in Nagel (1979a).

Nagel, T. (1978). 'Equality', reprinted in Nagel (1979a).

Nagel, T. (1979a). *Mortal Questions* (Cambridge: Cambridge University Press).

Nagel, T. (1979b). 'Subjective and Objective', in Nagel (1979a).

Nagel, T. (1986). *The View From Nowhere* (Oxford: Oxford University Press).

Nagel, T. (1991). *Equality and Partiality* (Oxford: Oxford University Press).

Narveson, J. (1967). 'Utilitarianism and New Generations', *Mind*, 76, 62–72.

Narveson, J. (1973). 'Moral Problems of Population', *The Monist*, 57, 62–86.

Narveson, J. (2000). 'Is There a Duty to Die?', in Humber and Almeder (2000).

Norman, R. (1996). 'Interfering with Nature', *Journal of Applied Philosophy*, 13 (1), 1–10.

Notestein, F. (1970). 'Zero Population Growth', reprinted in Kammeyer (1975).

Nuffield Council on Bioethics (2005). *The Ethics of Research Involving Animals* (London).

Nussbaum, M. C. (1986). *The Fragility of Goodnes. Luck and Ethics in Greek Tragedy and Philosophy* (Cambridge: Cambridge University Press).

Oakeshott, M. (1948). 'Rationalism in Politics', in Oakeshott, *Rationalism in Politics and Other Essays* (Indianapolis: Liberty Press, 1991).

Oderberg, D. (2000a). *Applied Ethics. A Non-Consequentialist Approach* (Oxford: Blackwell).

Oderberg, D. (2000b). *Moral Theory. A Non-Consequentialist Approach* (Oxford: Blackwell).

Olshansky, S. J., Carnes, B. A. and Cassel, C. (1990). 'In Search of Methuselah: Estimating the Upper Limits to Human Longevity', *Science*, 250, 634–40.

Olshansky, S. J., Carnes, B. A. and Désequelles, A. (2001). 'Prospects for Human Longevity', *Science*, 291, 1491–2.

Olshansky, S. J., Hayflick, L. and Carnes, B. A. (2002). 'Position Statement on Human Aging', *Journal of Gerontology: Biological Sciences*, 57A (8), B292–B297.
Olson, E. T. (1997). *The Human Animal. Personal Identity without Psychology* (Oxford: Oxford University Press).
O'Neill, O. (1996). 'Kant's Virtues', in Crisp (1996).
Overall, C. (2003). *Aging, Death and Human Longevity. A Philosophical Inquiry* (Berkley and LA: University of California Press).
Parens, E. (ed.) (1998). *Enhancing Human Traits. Ethical and Social Implications* (Washington, D.C.: Georgetown University Press).
Parfit, D. (1984). *Reasons and Persons* (Oxford: Clarendon).
Parfit, D. (1986). 'Overpopulation and the Quality of Life', in Singer (1986).
Perrett, R. W. (2000). 'Taking Life and the Argument from Potentiality', *Midwest Studies in Philosophy*, 24, 186–97.
Persson, I. (2003). 'Two Claims About Potential Human Beings', *Bioethics*, 17, 503–16.
Pijnenburg, M. A. M. and Leget, C. (2007). 'Who Wants to Live Forever? Three Arguments Against Extending the Human Lifespan', *Journal of Medical Ethics*, 33, 585–7.
Pitcher, G. (1984). 'The Misfortunes of the Dead', *American Philosophical Quarterly*, 21, 183–8.
Post, S. G. (1991). 'The Echo of Nuremberg: Nazi Data and Ethics', *Journal of Medical Ethics*, 17, 42–4.
Prichard, H. A. (1912). 'Does Moral Philosophy Rest on a Mistake?', reprinted in Prichard, *Moral Obligation*, W. D. Ross (ed.) (Oxford: Clarendon Press, 1949).
Priest, G. (2000). *Logic* (Oxford: Oxford University Press).
Pybus, E. M. (1982). '"Saints and Heroes"', *Philosophy*, 57, 193–9.
Rachels, J. (1975). 'Active and Passive Euthanasia', reprinted in Singer (1986).
Rainbolt, G. (2000). 'Perfect and Imperfect Obligations,' *Philosophical Studies*, 98, 233–56.
Rawls, J. (1971). *A Theory of Justice* (Oxford: Oxford University Press).
Raz, J. (1975). 'Permissions and Supererogation', *American Philosophical Quarterly*, 12, 161–8.
Raz, J. (2001). *Value, Respect and Attachment* (Cambridge: Cambridge University Press).
Regan, T. (1985). 'The Case for Animal Rights', reprinted in Baird and Rosenbaum (1991).
Reichlin, M. (1997). 'The Argument from Potential: A Reappraisal', *Bioethics*, 11, 1–23.
Rosenbaum, S. (1989). 'Epicurus and Annihilation', *Philosophical Quarterly*, 39, 81–90.
Ross, W. D. (1930). *The Right and the Good* (Indianapolis: Hackett).
Russell, B. (1905). 'On Denoting', *Mind*, 14, 479–93.
Sartre, J.-P. (1973). *Existentialism and Humanism*, Philip Mairet (trans.) (London: Methuen).
Scarre, G. (2007). *Death* (Stocksfield: Acumen).
Searle, J. R. (1958). 'Proper Names', *Mind*, 67, 166–73.
Sen, A (1980). 'Equality of What?' in S. McMurrin (ed.). *The Tanner Lectures on Human Values* (Salt Lake City: University of Utah Press) [http://www.tannerlectures.utah.edu/lectures/documents/sen80.pdf]

Singer, P. (1980). 'The Significance of Animal Suffering', reprinted in Baird and Rosenbaum (1991).

Singer, P. (ed.) (1986). *Applied Ethics* (Oxford: Oxford University Press).

Singer, P. (ed.) (1993a). *A Companion to Ethics* (Oxford: Blackwell).

Singer, P. (1993b). *Practical Ethics*, 2nd edn (Cambridge: Cambridge University Press).

Singer, P. (1995). *Animal Liberation*, 2nd edn (London: Pimilico).

Sprigge, T. (1968). 'Professor Narveson's Utilitarianism', *Inquiry*, 11, 332–48.

Strawson, G. (2004). 'Against Narrativity', *Ratio* (new series), 16, 428–52.

Strawson, P. F. (1950). 'On Referring', *Mind*, 59, 320–44.

Steinbock, B. (2006). 'The Morality of Killing Human Embryos', *Journal of Law, Medicine and Ethics*, 34, 26–34.

Steiner, G. (1989). *Heidegger* (Chicago: Chicago University Press).

Stirner, M. (1995). *The Ego and Its Own*, David Leopold (trans. and ed.) (Cambridge: Cambridge University Press).

Taurek, J. M. (1977). 'Should the Numbers Count?', *Philosophy and Public Affairs*, 6, 293–316.

Teichman, J. (1985). 'The Definition of *Person*', *Philosophy*, 60, 175–85.

Temkin, L. (1997). 'Rethinking the Good, Moral Ideals and the Nature of Practical Reasoning', in Jonathan Dancy (ed.). *Reading Parfit* (Oxford: Blackwell).

Tollefsen, C. (2001). 'Embryos, Individuals, and Persons: An Argument Against Embryo Creation and Research', *Journal of Applied Philosophy*, 18, 65–77.

Tooley, M. (1998a). 'Personhood', in Kuhse and Singer (1998).

Tooley, M. (1998b). 'Value, Obligation and the Asymmetry Question', *Bioethics*, 12, 111–24.

Tur, R. H. S. (2002). 'How Unlawful is "Euthanasia"?', *Journal of Applied Philosophy*, 19, 219–32.

Uniacke, S. (1994). *Permissible Killing. The Self-Defence Justification of Homicide* (Cambridge: Cambridge University Press).

Uniacke, S. (2004a). 'Harming and Wrongdoing: The Importance of Normative Context', in Timothy Chappell and David Oderberg (eds). *Human Values. New Essays on Ethics and Natural Law* (Basingstoke: Palgrave Macmillan).

Uniacke, S. (2004b). 'Is Life Sacred?', in Ben Rogers (ed.). *Is Nothing Sacred?* (London: Routledge).

Urmson, J. O. (1958). 'Saints and Heroes', reprinted in Joel Feinberg (ed.) *Moral Concepts* (Oxford: Oxford University Press, 1969).

Urmson, J. O. (1988). *Aristotle's Ethics* (Oxford: Blackwell).

Warnock, M. (1985). *A Question of Life. The Warnock Report on Human Fertilisation and Embryology* (Oxford: Blackwell).

Warren, M. A. (1987). 'Difficulties with the Strong Animal Rights Position', reprinted in Baird and Rosenbaum (1991).

Warren, M. A. (1997). 'On the Moral and Legal Status of Abortion', in LaFollette (1997).

Wiggins, D. (1975). 'Deliberation and Practical Reason', reprinted in Wiggins (2002).

Wiggins, D. (1979). 'The Concern to Survive', reprinted in Wiggins (2002).

Wiggins, D. (1987). 'Claims of Need', reprinted in Wiggins (2002).

Wiggins, D. (1998). 'The Right and the Good and W.D. Ross's Criticism of Consequentialism', *Utilitas*, 10, 261–80.

Wiggins, D. (2002). *Needs, Values, Truth*, 3rd edn (Oxford: Clarendon).

Williams, A. (1997). 'Intergenerational Equity: An Exploration of the "Fair Innings" Argument', *Health Economics*, 6, 117–32.

Williams, B. (1956). 'Personal Identity and Individuation', reprinted in Williams (1973c).

Williams, B. (1962). 'The Idea of Equality', reprinted in Williams (1973c).

Williams, B. (1965). 'Ethical Consistency', reprinted in Williams (1973c).

Williams, B. (1973a). 'A Critique of Utilitarianism', in J. J. C. Smart and Bernard Williams, *Utilitarianism: For and Against* (Cambridge: Cambridge University Press).

Williams, B. (1973b). 'The Makropulos Case: Reflections on the Tedium of Immortality', in Williams (1973c).

Williams, B. (1973c). *Problems of the Self. Philosophical Papers 1956–72* (Cambridge: Cambridge University Press).

Williams, B. (1976a). 'Moral Luck', reprinted in Williams (1981).

Williams, B. (1976b). 'Persons, Character and Morality', reprinted in Williams (1981).

Williams, B. (1980). 'Internal and External Reasons', reprinted in Williams (1981).

Williams, B. (1981). *Moral Luck. Philosophical Papers 1973–1980* (Cambridge: Cambridge University Press).

Williams, B. (1988a). 'What Does Intuitionsim Imply?', reprinted in Williams (1995).

Williams, B. (1988b). 'Which Slopes Are Slippery', reprinted in Williams (1995).

Williams, B. (1993). *Ethics and the Limits of Philosophy* (amended edn) (London: Fontana Press).

Williams, B. (1995). *Making Sense of Humanity and Other Philosophical Papers 1982–1993* (Cambridge: Cambridge University Press).

Williams, B. (2006). 'The Human Prejudice', in Williams, *Philosophy as a Humanistic Discipline*, A. W. Moore (ed.) (Princeton and Oxford: Princeton University Press).

Williams, G. C. (1957). 'Pleiotropy, Natural Selection, and the Evolution of Senescence', *Evolution*, 11, 398–411.

Winch, P. (1964). 'Understanding a Primitive Society', reprinted in Bryan R. Wilson (ed.). *Rationality* (Oxford: Blackwell, 1970).

Wittgenstein, L (1967). *Philosophical Investigations*, 3rd edn, G. E. M. Anscombe (trans.) (Oxford: Blackwell).

Wolf, S. (1982). 'Moral Saints', *Journal of Philosophy*, 79, 419–39.

Wolheim, R. (1956). 'Equality', *Proceedings of the Aristotelian Society*, 56, 281–300.

Wollheim, R. (1984). *The Thread of Life* (New Haven and London: Yale University Press).

Woodger, J. H. (1945). 'On Biological Transformations', in W. E. Le Gros Clark and P. B. Medawar (eds). *Essays on Growth and Form* (Oxford: Clarendon Press).

Index

activities and attachments, 43, 46, 87,
    118, 125, 140, 144
  character, 118–19, 132
  convictions, 57–9
  cultural background, 18, 23–24, 39
  defined, 17
  equality, 126–7
  limiting longevity, 85–6, 135–6
  misfortune of death, 29–30, 38
  personal identity, 27
  self-concern, 114–15, 116, 122–4,
    133
  suicide, 107–9
  tedium of immortality, 45–8
  value of life, 17, 18, 39, 40, 47–48,
    74, 76, 108, 118, 122, 123, 126,
    127, 134
ageing,
  as a disease, 8–10, 12, 17, 92–3,
    140, 146 n. 10
  defined, 3
  rate of, 3, 145 n. 5
ageing process, 3, 4, 10–12, 45, 84, 87,
    93, 94, 99, 135, 139, 140, 143
  and associated diseases and
    disorders, 8, 11, 12, 15, 84, 87,
    93, 99, 135, 137, 140
ageing, theories for,
  evolutionary, 5–6
  programmed, 4–5,
  wear and tear, 3–4,
  see also, Disposable Soma Theory
Alderman, H., 22
animals, 4, 5, 8, 51–60, 62, 63, 72
    125, 140–1
  capacity to suffer, 52
  equality of interests, 52–3, 55, 57
  rights of, 53–5, 56
  speciesism, 53, 55–6, 63
  suffering, 51
  see also, Integrity, P. Singer,
    T. Regan
Anscombe, G.E.M., 134, 150 n. 9

Aristotle, 24, 65–6, 106, 128, 129,
    131, 132, 153 n. 24
assisted suicide, 88, 100–1, 103, 112,
    153 n. 14
asymmetry,
  between life and death, 36
  between twins and non-twins, 70
  of non-existence, 32, 40–4, 149
    n. 25
  of obligations when having
    children, 78–9
Austad, S., 5, 145 n. 5
Average Principle, 79, 80, 81, 82, 83

Battin, M., 152 n. 8
Bauman, Z., 16, 23
Belshaw, C., 36
biological warranty period, 2–4, 8, 12,
    15, 25, 91–4
birth rate, 73, 76–8, 84, 85–6, 88, 134,
    142
Boorse, C., 9, 146 n. 11, 146 n. 16
Bradley, B., 35, 36
Broome, J., 151 n. 3, 151 n. 6

Calment, J., 2
Cambridge change, 36–7
Camus, A., 17
Čapek, K., 45, 47, 149 n. 27
Carnes, B., 2–3, 8, 91, 145 n. 3
Carter, A., 80–3, 151 n. 10
categorical desires, 29–31, 35, 39, 40,
    41–2, 45–8, 97, 100, 140, 148 n. 6
character, 45–7, 118–20, 122, 123,
    124, 127, 131–2, 133–4
charity, 109, 133
clearing, see cultural background
Coale, A., 77
commitments, 18–19, 23, 39, 57–8,
    72, 122, 124, 126–7, 141
Cottingham, J., 117–19, 155 n. 8, 155
    n. 22, 155 n. 24
Crisp, R., 129, 131

cultural background, 18–19, 23–5,
    38–40, 75, 133, 142, 147 n. 30,
    147 n. 31, 151 n. 1

Dancy, J., 130, 131, 155 n. 20
Daniels, N., 9, 13, 90, 115–16, 152 n. 4
death,
    authenticity, 17, 107
    extrinsic causes of, 3, 4, 6, 7, 8,
        49, 92
death, misfortune of, 28–44
    arguments explaining, 28–31
    objections to, 31–2
    over-determining, 37–40
    phenomenology of, 32–4
    referential failure, 34
    timing of, 34–7
    see also, T. Nagel, B. Williams
Disposable Soma Theory, 6–8, 9, 11,
    12, 16, 17, 39, 77, 91–2
Doctrine of Double Effect, 103, 104,
    153 n. 18
Dreyfus, H., 18, 22, 23, 147 n. 30
duty to die, 96–8, 100, 107–8, 113,
    152 n. 8
duties, 104, 106, 107, 121, 133, 153
    n. 19
    against killing, 109–11
    conflicts of, 106
    perfect/imperfect distinction, 110
Dworkin, R., 102, 108

Eberl, J., 67
Eliot, T.S., 19
Elitism,
    Mere Addition Paradox, 82
    partialism, 119, 122–4
Elliott, C., 14
embryos, 60-72, 126, 154 n. 1
    as a human being, 62–3
    as a Person, 62–3
    creating embryos for experiments,
        71–2
    interests of, 63–4
    see also, potentiality
enhancement, 12–15, 92, 139
    consequences of, 14–15
    defined, 12–13
    means of achieving, 14, 50
    promotion of values, 13–14

Epicurus, 31–2, 44, 148 n. 10, 148
    n. 11
equality, 73, 81–4, 86, 89, 90–1, 95
    96–7, 117–19, 120, 134–8, 141–2,
    151 n. 9, 152 n. 8
    see also, animals, Principle of
    Equality, Principle of Equal
    Respect
euthanasia, 88, 100–3, 107, 109–12,
    152 n. 11, 153 n. 14, 154 n. 29

Feinberg, J., 104–7, 153 n. 21, 153
    n. 23
fair innings argument, 88–92, 94–5,
    98, 138, 152 n. 5
Finnis, J. 102, 153 n. 20, 154 n. 25
Five Dimensions Theory, 83–6, 135,
    136, 142
flourishing, *see* the good life
Foot, P., 103, 150 n. 9, 154 n. 27
fragility, 14, 24–5, 47–8, 47–8, 72,
    108, 141, 143–4
freedom (and autonomy), 17, 19, 23,
    107–9, 114
Freud, S., 15–16, 146 n. 19
Fukuyama, F., 87
functions, 4, 6–7, 26, 28, 146 n. 11,
    146 n. 16
    normal, species typical, 9, 12–13,
        98, 135, 137, 140
    practical, 9–10, 12, 92–3, 137, 140
    teleological, 9–10, 12, 92–3, 137
fundamental moral requirements,
    114, 115, 119–20, 123–4,
    126–28, 133–4, 155 n. 10

Garland, R., 100, 152 n. 11
Gems, D. 151 n. 12
Glover, J., 48
good life, the, 23–25, 31, 32, 48, 50,
    58, 60, 63, 72, 74–8, 80, 83–6,
    88, 90, 93, 95, 97, 118–19, 122,
    124, 132–8, 140–44, 151 n. 1,
    155 n. 22
Gorer, G., 16
Gruman, G., 148 n. 11
Guarente, L., 9
Haber, C., 146 n. 10, 146 n. 13
Hardwig, J., 96, 107, 152 n. 8
Hare, R.M., 75–6

Harris, J., 89, 151 n. 12, 152 n. 2, 152 n. 11
Hart, H.L.A., 103, 150 n. 7
Hayflick, L., 5, 9
healthcare, 13, 73, 88–91, 92, 94–9, 112, 115–6, 135, 137–8, 142, 152 n. 5, 152 n. 7
  definition of health, 12–13
  genuine medical needs, 13
Hegel, G.W.F., 16
Heidegger, M., 17–24, 72, 107, 132, 144, 146 n. 23, 147 n. 25, 147 n. 28, 147 n. 30, 147 n. 31, 148 n. 4, 154 n. 26
Herodotus, 100, 152 n. 11
Holland, A., 67, 68
human activities, 92–4, 99, 116, 140, 148 n. 11, 152 n. 7
  and nature, 10
Hursthouse, R., 56, 150 n. 8

Ideal Observer, 55, 117–18, 120
immortality, 11, 32, 145 n. 5, 149 n. 23
  tedium of, 44–9, 85–6, 88, 109, 110, 137, 143, 154 n. 26
impartialism, 95, 117–3, 120–8, 137, 155 n. 24
integrity, 56–60, 72, 108, 140–1
intelligent adult (example), 33–4, 38, 43, 46–7, 148 n. 13

Jim and the Indians (example), 57, 59, 130, 150 n. 9
Jonas, H., 19–21, 74, 144
justice, 76, 88, 90, 102, 106, 122–4, 133, 151 n. 9, 152 n. 2, 154 n. 2, 154 n. 24, 154 n. 27, 155 n. 8, 155 n. 22, 155 n. 24

Kaufman, F., 148 n. 7
Kenyon, C., 9
King, C.D., 9
Kirkwood, T.B.L., 4, 5, 6–8, 15, 16, 145 n. 2, 145 n. 5
Kripke, S., 34, 42, 149 n. 26
Kuhn, T., 77, 147 n. 30

life span,
  average expected, 1–3, 8, 90, 145 n. 2
  biblical account of, 1, 145 n. 1

incidental increase of, 10–11, 12
intentional increase of, 11, 12
maximum, 2
species typical, 4, 5, 6, 7, 8, 91, 92
Little, M., 13–14
living well, *see* the good life
longevity, *see* the good life, life span, natural life span, normal life span, quality of life, values
Lucretius, 31, 32, 40, 44, 148 n. 15

Malthus, T., 74
maximum life span (imposed), 96–9, 100, 101, 111
  see also, normal life span
McDowell, J., 132
McMahan, J., 37, 147 n. 1, 149 n. 19, 149 n. 22, 149 n. 28
Medawar, P., 6
Mere Addition Paradox, 81–2
minimum birth rate, 76–8, 84–6, 142
moral perception, 106–7, 111, 128–31, 132, 153 n. 24
moral saint, 122, 132
mortality rate, 3, 11, 73, 76, 84, 86
  of infants, 2

Nagel, T., 26, 28–9, 30, 33, 35–6, 37, 38, 40, 44, 47, 117, 148 n. 12, 148 n. 15, 149 n. 21, 149 n. 25, 154 n, 4
Narveson, J., 79, 96, 97, 151 n. 3, 151 n. 5
natural law, 106, 154 n. 25
natural life span, 91–4, 95, 138, 144, 152 n. 7
Natural Life Span View, 92–3
nature, 10, 16
normal life span,
  as a background feature, 38–9, 40
  as a fair innings, 88, 89–91, 92, 94
  imposed, 96, 98–9, 100, 111–12, 115, 137–8, 141–2
Normative Claim, 114, 115, 123, 125, 127

Oakeshott, M., 147 n.25
objective view, 33–4, 35, 37, 39–40, 44, 55, 81, 117, 137, 148 n. 12, 154 n. 4

Oderberg, D., 102, 153 n. 15
Olshansky, S.J., 33, 11, 145 n. 2
optimal population levels
    nominal, 75, 83, 151 n. 1
    real, 75–6, 78, 86
overpopulation, 4–5, 73-86, 134, 135,
    142, 151 n. 12
    nominal, 75
    possibility of, 74
    real, 75, 85, 135

Parfit, D., 27, 41, 42, 43, 79–84, 151
    n. 11
partialism, 120–24, 125–8, 132, 155
    n. 8, 155 n. 24
particularism, 56, 59, 106, 109, 121,
    126, 128–31, 133, 153 n. 24
Perfectionist Theory, 82–3
Perret, R., 70
personal identity, 26–8, 35, 42–4, 45,
    147 n. 1, 149 n. 26
Plato, 24
potentiality, 64–71
    active/passive distinction, 65–6,
    70–1
    definitions, 64–5
    extra-embryonic material, 67–8
    twinning, 68–70
Principle of Equal Respect, 125–7,
    128, 132, 134, 138
Principle of Equality, 113–5, 116,
    117, 119, 123, 124, 125
Prichard, H.A., 131, 155 n. 14
Prudential Life Span Account,
    115–17, 118, 154 n. 2

quality of health, 9–11, 12, 25,
    31, 50, 60, 71, 72, 84, 87–88, 93,
    94–5, 96, 98–100, 108–9, 112,
    116, 135–7, 140–4,
    146 n. 13
quality of life, 11, 25, 48, 74–6, 79,
    80–4, 85, 112, 135–36, 142–43
    see also, the good life

Rawls, J., 82, 117, 118, 138, 151
    n. 9
Raz, J., 114, 126, 155 n. 10

Regan, T., 51, 53–5, 56, 58, 64, 80,
    125
Reichlin, M., 66
respect, 71, 72, 120
    see also, Principle of Equal Respect
right to life, 100, 103–7, 109–111,
    conflicts between, 104–107
Rosenbaum, S., 148 n. 10
Ross, W.D., 104, 106, 128, 129, 130,
    131, 153 n. 19, 153 n. 20, 155
    n. 14
Russell, B., 34

sanctity of life, 101–3, 104, 105, 107,
    108, 111, 153 n. 15
Sartre, J.-P., 17, 23
Scarre, G., 35, 37, 41
Searle, J., 34
Sen, A., 152 n. 4
Singer, P., 51–3, 55, 56, 58, 59, 115,
    125, 150 n. 7, 151 n. 10
Sprigge, T., 151 n. 5
stem cell therapies, 50, 60–1, 67, 71,
    72
Stirner, M., 122
Strawson, P.F., 34
subjective view, 33–34, 39–40, 75, 86,
    88, 109, 117, 125, 137, 148 n.
    12, 154 n. 4
suicide, 88, 96, 97, 100–3, 107–9,
    110, 112
supererogation, 97, 114, 121, 132

Taurek, J.M., 150 n. 9
technological understanding of
    Being, 21–5, 31, 32, 48–9, 60, 72,
    141, 144, 147 n. 31
technology,
    and values, 15, 19–25, 139, 143–4
    determining objectives, 19–21,
    pre-modern/modern distinction,
    20–1, 22, 24
    unevenness of, 87–8, 100
    overpopulation, 74, 95, 142–3
    see also, enhancement,
    technological understanding of
    Being
Total Principle, 79, 80, 81, 82, 83

Uniacke, S., 148 n. 8
Urmson, J.O., 97

value of life, 23, 29, 30, 49, 52, 54,
  64, 71, 76, 101–2, 103, 104, 107,
  108, 109, 110, 111, 113–14, 115,
  117–19, 120, 125–7, 132, 134
values,
  and the cultural background, 19,
    23–4
  of children, 76
  of longevity, 11, 25, 31, 32, 44, 48,
    60, 87, 136–8, 140
  trade-offs between, 75, 83–4, 85, 142
  see also, activities and attachments,
    enhancement, health,
    technology

virtues, 118, 119–20, 133–4, 141, 155
  n. 24
vulnerability, *see* fragility

Weismann, A., 6, 15
Wiggins, D., 106
Williams, A., 89, 90, 98
Williams, B., 17, 29–30, 31, 38, 45–8,
  55–7, 58, 62, 85, 117, 125, 134,
  148 n. 6, 149 n. 28, 150 n. 8,
  150 n. 10, 155 n. 21
Williams, G., 6
Wittgenstein, L., 18, 131, 133, 155
  n. 21
Wolf, S., 122
Wollheim, R., 29, 149 n. 23
Woodger, H.G., 62